Intellectual Entertainments

The symposium by Roger Vieillard

Anthem Studies in Wittgenstein publishes new and classic works on
Wittgenstein and Wittgensteinian philosophy. This book series aims
to bring Wittgenstein's thought into the mainstream by highlighting its
relevance to 21st century concerns. Titles include original monographs,
themed edited volumes, forgotten classics, biographical works and books
intended to introduce Wittgenstein to the general public. The series is
published in association with the British Wittgenstein Society.
Anthem Studies in Wittgenstein sets out to put in place whatever measures
may emerge as necessary in order to carry out the editorial selection
process purely on merit and to counter bias on the basis of gender, race,
ethnicity, religion, sexual orientation and other characteristics protected
by law. These measures include subscribing to the British Philosophical
Association/Society for Women in Philosophy (UK) Good Practice Scheme.

Series Editor

Constantine Sandis – University of Hertfordshire, UK

Intellectual Entertainments

Eight Dialogues on Mind, Consciousness and Thought

P. M. S. Hacker

ANTHEM PRESS

Anthem Press
An imprint of Wimbledon Publishing Company
www.anthempress.com

This edition first published in UK and USA 2020
by ANTHEM PRESS
75–76 Blackfriars Road, London SE1 8HA, UK
or PO Box 9779, London SW19 7ZG, UK
and
244 Madison Ave #116, New York, NY 10016, USA

First published in the UK and USA by Anthem Press 2019

British Library Cataloguing-in-Publication Data
A catalogue record for this book is available from the British Library.

Library of Congress Control Number: 2020944638

ISBN-13: 978-1-78527-555-5 (Pbk)
ISBN-10: 1-78527-555-0 (Pbk)

This title is also available as an e-book.

For
Jonathan, Adam and Jocelyn

CONTENTS

PREFACE

The dialogue is one of the oldest forms in which to present philosophical ideas and the lively clash of philosophical disputation. On the one hand, it allows the author, if he so pleases, to hide behind the characters he has created. More importantly, it permits him to present ideas with which he does not agree, with all the commitment and passion that their adherents feel. And it enables him to display how difficult it is to uproot received and deeply tempting ideas. On the other hand, it makes it possible for the audience to follow a lively debate, to hear the different views that have been advanced by various thinkers and to come to their own conclusion. Its primary function, however, is to stimulate thought and discussion among its readers and, if performed as a reading or dramatization, its listeners, by means of an imaginary discussion between protagonists advancing different views.

The following eight dialogues were written, as their title intimates, as intellectual entertainment. They were not written primarily with an academic readership in mind. On the contrary, they were written in order to go over the heads of professional academics and to reach the wider audience of those with an interest and curiosity in matters intellectual. There are, after all, few intelligent people who have not wondered about the nature of the mind and about the relation of the mind to the body, and who have not, from time to time, paused to reflect on the question of whether the mind is the same as the self, or whether the mind is identical with the brain. So the two dialogues on these subjects will, I hope, provide intellectual entertainment, and induce further reflection. In the present intellectual milieu, in which we are bombarded in press, radio and television with over-hasty news from cognitive neuroscience, it is difficult, even for the most thoughtful people, not to succumb to the enchantments of mysteries. For we are assailed by academics – in philosophy, psychology and cognitive neuroscience – eager to persuade us that consciousness is profoundly mysterious, and, moreover, that it is the last barrier to a fully scientific conception of the universe. The two dialogues on consciousness, its nature and its various forms were designed to provide food for thought and to show that the mysteries of consciousness are no more than mystifications in colourful and alluring pseudo-scientific clothing. As one of my characters remarks, 'There are no mysteries' could be the motto written on the coat of arms of Philosophy. The third major theme – the nature of thought, and hence too the relationship between thought and language, the question of what we think in and of whether neonates and animals can think – is a collection of inter-related conundrums that have perennially puzzled and bewildered thoughtful people. I trust that the two dialogues on these subjects will both amuse and provoke further thought.

Although these dialogues were not written for academic purposes, undergraduates studying philosophy may profit from reading them. They will not help students to pass their examinations, but they may stimulate them to think for themselves and to challenge academic orthodoxy. They may also provide a good basis for discussions in classrooms at schools where philosophy is taught.

The protagonists in the eight dialogues often include august dead philosophers. Socrates, Aristotle, Descartes, Locke, Frege, Wittgenstein, Alan White and Peter Strawson all appear in one or more of the dialogues. With the exception of Socrates and White, the views given by them in my dialogues are for the most part theirs, and their words are often quotations or paraphrased quotations from their writings and letters. I have endeavoured not to put into their mouths any judgement that they would not, to the best of my knowledge, have accepted. On the other hand, I have not aimed at perfectly accurate representations of their views, only of their points of view – for these are imaginary dialogues, not philosophical records, and the protagonists are shades or shadows of the real philosophers. Nevertheless, I have also tried, as best I could, to emulate their manner of writing, their orthography, the forms of their 'white noise' in correspondence, and their styles of abuse, condescension and self-reference. My imaginary characters serve various purposes – to represent a standard viewpoint of one kind or another, to be a foil for another character and to fulfil the role of an intelligent but philosophically naïve person. In the introductions to the different dialogues or pairs of dialogues, I have commented on some of the imaginary characters more specifically.

I have peppered each of the dialogues with endnotes. These are not meant to be consulted while reading the dialogue, for that would disrupt the flow of conversation. They are there for two purposes. Notes to remarks by specific historical figures serve to guide the reader to the texts from which some of the statements are derived. Notes to remarks made by my imaginary characters, especially my cognitive neuroscientists and wayward philosophers, are given to assure the reader that some of the weird and wonderful things said by my imaginary characters are not the products of my fevered mind, but are actually the views of distinguished men of science (and philosophers) of recent decades.

In each section, at the end of the first dialogue, I have appended a very brief reading list for those who might wish to pursue matters further. I have referred to writings of mine that elaborate some of the views cursorily discussed in the dialogue, and to two or three works of others that I have found particularly helpful, interesting or relevant to the dialogue.

ACKNOWLEDGEMENTS

I began writing these dialogues largely for my own amusement and the entertainment of my friends. They joined in the fun. They gave me much encouragement, as well as copious comments and corrections that have improved the text. I am indebted to Hanoch Ben-Yami, John Cottingham, Parashkev Nachev, Hans Oberdiek, Herman Philipse, the late Dan Robinson, Amit Saad and David Wiggins for reading and commenting upon the draft dialogues. I am grateful to Keith Thomas for corrections to my seventeenth-century English prose, and to my son Jonathan Hacker for his advice on the dialogue form.

When I discovered that I could actually read the dialogues out aloud and produce five different voices for the five characters with moderate competence, I began to give the occasional public performance for the entertainment of a larger audience. I am grateful to members of the audiences at Oxford Brookes University, at the University of Utrecht, at University College, London, at the Stuart Hampshire Philosophical Society of Rugby School, at the Royal Institute of Philosophy in London, at Oxford University, at the University of Kent, at the Ashmolean Museum, Oxford, at the University of Thessaloniki and at the Hebrew University of Jerusalem. Their questions enabled me to sharpen my dialogues and stimulated me to add further exchanges in them.

The dialogue 'Can Different People Have the Same Pain?' (an ancestor of the Eighth Dialogue, with the same name) was published in the *Tomsk Journal of Philosophy* in Russian translation in 2012. 'A Dialogue on Secondary Qualities' (an ancestor of the Fifth Dialogue, 'On the Objectivity or Subjectivity of Perceptual Qualities') was published in *Iyyun –The Jerusalem Philosophical Quarterly*, 63 (January 2014) in an issue dedicated to the memory of Edna Ullmann-Margalit. The dialogue 'The Nature of the Mind' (First Dialogue) was published in Spanish translation in *Dokos* (2014) and in English in Anthony O'Hear (ed.), *Mind, Self and Person*, Royal Institute of Philosophy supplement, 76 (2015). The dialogue 'The Mind and the Body' (Second Dialogue) was published in *Philosophy* 89 (October 2014). The two dialogues 'Thought' (Sixth Dialogue) and 'Thought and Language' (Seventh Dialogue) were published in *Philosophy*, 92 (2017).

SECTION 1

TWO DIALOGUES ON
MIND AND BODY

INTRODUCTION

We are human beings – living animals of the species *homo sapiens* – with a distinctive range of abilities. We have a body; but it is far from evident that we are identical with the body we have. We have a mind; but it is equally unclear whether we are identical with the mind we have. We are also commonly said to have a soul and indeed a self. But if we *have* all these things, *who* and *what* are we that have them? And *what* exactly is it that we have? Is the soul identical with the mind or is it distinct from it? Is the divide between mind and body the same as the divide between body and soul? Sensual appetites are traditionally assigned to the body or the flesh, not to the soul. But desire is commonly assigned to the mind not to the body. And remorse and shame are commonly assigned to the soul, not to the mind. So surely the mind and the soul are distinct. I am, no doubt, myself; but am I my *self*? And what and where is *a self*? Small wonder that these questions have puzzled philosophers since the dawn of philosophy, bewildered theologians, baffled doctors and perplexed psychologists and neuroscientists.

To be sure, being living animals, we *are* bodies – animate spatio-temporal beings consisting of flesh and blood. But how can something that *is* a body also *have* a body? Surely one cannot *have* what one *is*? We have a mind, but what is it that human beings *have* when they have a mind? And how can human beings be identical with something they *have*? I am myself, but I surely don't *have* myself! A human body is a space occupant, but what is a human mind? Is it material or immaterial? Is the mind the brain? – if so, then a mind weighs three pounds and is seven inches tall! Or is the mind something immaterial? One might say, with some plausibility, that a human being is a combination of mind and body. But if that is right, then what sort of combination is involved? How can something immaterial be 'combined' with a body? And how can an immaterial mind or soul *interact* with a material body? And if they are combined, are they separable? Can the mind, or soul, survive the death of the body? But then, is it *bodies* that die and not human beings?

These questions are deep. They reach down to the very roots of the forms of our thought about ourselves. Different kinds of answer have been ventured to them throughout the history of philosophical reflection. Plato advanced a dualist conception of the relationship between the mind and the body, conceiving of them as two separate things temporarily united in the course of human life. The mind or soul, he thought, existed before its incarnation in the human body, and survives death. Aristotle advanced a monist conception – the body and the mind constitute a unity of matter and form, and to ask whether a human being is one thing or two is like asking whether a piece of wax and its shape are one thing or two. The mind, he argued, is constituted by the set of distinctive powers of intellect and rational will possessed by human beings. And the rest, one might say with only some exaggeration, is footnotes.

Platonic dualism was congenial to Christianity and was adapted to Christian doctrines by Augustine. Augustine in turn was a major influence on the father of early modern philosophy – Descartes. Cartesian dualism has been an inspiration and has posed a challenge to all philosophers since the seventeenth century. It is most striking that a majority of contemporary philosophers and virtually all contemporary cognitive neuroscientists reject Cartesian dualism, advancing the view that the mind is the brain. But they remain in the shadow of Descartes in as much as they unwittingly hold that the fundamental relation between the brain and the body is the same as the relation between the Cartesian mind and the body. But there is much more awry with Cartesian dualism than the misconception that the mind is an immaterial substance. Simply to reject the idea that the mind is an immaterial, spiritual, substance distinct from the body is merely the beginning of wisdom. *Much* more in the Cartesian (and empiricists') conception of the mind and of the relation between mind and body has to be scrutinized, anatomized and rejected before we can emerge from the baleful shadow of Descartes.

Aristotle's major influence was upon the philosophy of the High Middle Ages, and was adapted to the demands of Christianity by the noble endeavours of Thomas Aquinas. His influence, however, waned with the rise of renaissance neo-Platonism, and it was eclipsed by the rise of modern science in the seventeenth century. Aristotelian physics was shown definitively to be wrong by Kepler and Galileo. His philosophy of mind was unthinkingly swept aside together with his physics and then completely displaced by Cartesian philosophy of mind.

It is a striking fact that Ludwig Wittgenstein, the greatest philosopher of the twentieth century, both demolished Cartesian philosophy of mind and its various degenerate offshoots and advanced a novel and indeed revolutionary philosophy of mind that bears remarkable affinities to the Aristotelian one, even though he never read a word of Aristotle.

The following two dialogues on the nature of the mind and on the mind/ body relation discuss many of the most important and puzzling questions on these subjects. Since they include among the protagonists Aristotle, Descartes, Peter Strawson and Alan White, they are set in Elysium. With the exception of Alan White, the views expressed by them are, for the most part, their own, and often the words they utter are quotations or paraphrases of quotations. A third participant is Frank Craik, an imaginary American neuroscientist, who presents a Galtonian picture of views advanced by numerous eminent neuroscientists of recent decades. Richard, an Oxonian figure of the 1950s, represents the style and strength of the golden age of Oxford philosophy. Jill is both a sounding board and a foil for the others.

First Dialogue

ON THE NATURE OF THE MIND

Protagonists:

Richard: a middle-aged Oxford philosopher of the mid-twentieth century, dressed in cavalry twill slacks, waistcoat and tie, and well-cut jacket.

Jill: a philosopher in her early thirties, dressed in an elegant but informal manner.

Frank Craik: a contemporary American neuroscientist, casually dressed in jeans and pullover, with open-necked shirt. He has a marked American accent.

Descartes: in sombre Dutch mid-seventeenth century dress. Speaks with a thick French accent.

Aristotle: in Greek dress – himation and sandals.

The setting is a garden in Elysium. The sun is shining. A rich verdant lawn is surrounded by flower beds and flowering bushes, with a grove of magnificent trees behind. Beyond, there is a large lake and in the distance, high mountains. Five comfortable garden chairs are placed in the shade of some trees. There is a low table on which are placed a wine decanter and glasses, three of which are half full. Richard, Jill and Frank are deep in discussion.

Richard: But you must admit that it is very puzzling that we speak of *having* a mind and *having* a body. I mean, if I have a mind and also have a body, then who and what am I that has these two things?

Jill: Well, it seems obvious enough. After all, you just said '*I* have a mind' and '*I* have a body'. It is you, the 'I', the Ego, the Self, that has a mind, on the one hand, and a body, on the other.

Richard: But, Jill, what on earth is this 'I' or 'Ego' or 'self'? Surely I'm a human being.

Frank: Sure. And if you're a human being, then you can't be an Ego or Self. Unless human beings are selves.

Jill: All right. But then I surely *have* an Ego or Self. Human beings *have* selves.

Frank: No, no. Do I have a self? I've never come across it! I'm sure I'm sometimes selfish, but that doesn't mean that I have a self. And as for an Ego, that's just a fancy way of saying that I have an 'I'. It may sound better in Latin, but it's just baloney. Look, talking of *an 'I'* is just plain ungrammatical. I mean, y'don't talk of the you, the she or the it. Well, it's just as ungrammatical to talk of an I, of the I, or of my I.

Richard: [*chuckles*] Oh, my eye!

Jill: [*a little hotly*] All right. I grant you the ungrammaticality. Perhaps all this talk of '*an* "I"' and '*the* "I"' is ill-advised. But it doesn't follow that there is no such thing as a self, does it? After all, we speak, perfectly intelligibly, of our better self, and Polonius advised Laertes 'to thine own self be true' – you can't say that that's baloney, Frank.

Richard: [*pouring oil on troubled waters*] Take it slowly. We really need some clarity here … No one is going to quarrel with the statement that we have a mind and that we have a body. Some people want to insist, as Jill does, that we also have a self, and others like you, Frank, disagree. Let's shelve the disagreement for a moment and try to let some light in. In the first place, who is it that has a mind, and a body – and perhaps also a self?

Jill: Well, … It's me, *this* living human being.

Richard: So it's human beings who have minds, and have bodies, and have perhaps selves. So we're human beings, and we possess a mind, possess a body, maybe also possess a self. What about the soul? Do you also possess a soul, Jill?

Jill: Well, I'm not sure what to say. It's starting to look like an excess of possessions. [*She laughs*]

Frank: The soul is just pre-scientific mythology. Look, you, your joys and your sorrows, your memories and your ambitions, your sense of personal identity and free will, are in fact no more than the behaviour of a vast assembly of nerve cells and their associated molecules.[1]

Richard: That's a bit quick, Frank. In the first place, we aren't behaviour. And our joys and sorrows, our memories and ambitions, are not behaviour either, although they are *manifest* in behaviour. But they're manifest in *our* behaviour, not in the behaviour of our nerve cells. Secondly, we're flesh and blood – living animals constituted of a vast array of different kinds of cells. And, like all other organisms, we are also constituted of a variety of chemical elements, variously combined to form hugely complex molecules. But we are not identical with the stuff of which we're made, any more than we're identical with the ever changing assemblage of cells of which we consist.

Frank: Why not? Why aren't we identical with the matter – the material stuff – of which we're made?

Richard: The natural replacement over time of the cells of which a living organism is constituted does not change the identity of the organism. These mighty trees [*he waves at the trees beyond the garden*] are the same organisms as the little seedlings from which they grew, are they not? But neither the matter of which they are made nor the cells of which they are constituted are the same.

Frank: OK … Yeah … I can see that. I'm not a philosopher, and I'm not sure how to respond to your point. But it sure doesn't follow that you're a soul and that you're identical with your soul. That's just religious mythology. Y'don't *have* a soul. Souls don't exist.

Jill: And yet we do speak perfectly intelligibly of someone's being a soul in torment or of being a gentle soul.

Richard: And we also speak of someone losing their soul and of selling their soul, to the devil or to the company store, as the case may be. So, on your view, Jill, is the soul something we *are* or something we *have*?

Jill: It looks as if we both have a soul and are a soul. But that *is* paradoxical. I mean the owner cannot be identical with what she owns, can she? This is very odd. How is the soul related to the mind? And how is it related to the self? And how are the mind, the soul and the self related to the body one has? Does my mind belong to my body?

Richard: What would a body do with a mind? And if your body turned to stone, Jill, would your soul then belong to the stone statue?

Jill: Oh! … All right. So does my body belong to my mind?

Richard: And not to you? Is it really your mind that has a body? If that is right, then who on earth is it that has a mind?

Jill: Well, it's obviously me — *I* have a mind.

Frank: OK, OK. But now we're just going round in circles.

[*Descartes strolls out of the trees*]

Descartes: *Bonjour, mes amis.* I could not help hearing you conversing as I was taking my afternoon stroll. The topic about which you are discoursing is a deep and important one. Your ardour is *admirable*, although your reasoning may be questioned.

Richard: Well, please do join us here, sir. This lady is Jill.

Descartes: [*bows and doffs his hat*] *C'est un honneur et un plaisir, Madame.*

Richard: This is Frank, a brain scientist [*Descartes smiles and raises his eyebrows*] and my name is Richard. I'm a philosopher.

Descartes: [*bows*] *Messieurs.*

Richard: We should be delighted if you were to join us, sir. Do sit down. Would you care for a glass of wine? [*He pours a glass of wine and hands it to Descartes*]

Descartes: *Merci, merci.* [*He takes a deep drink*] Ah, *très bien.* It would be most *agréable* to sit here under the trees and join your debate. I gather from what I heard that you are concerned with the relation between the mind and the body, *n'est-ce pas?*

Jill: Yes, that's right. We were trying to get clear about what exactly we are, whether we're minds or egos or selves.

Richard: The question, I think, is what a human being is. I mean we speak of having a mind, and of having a body. And it seems as if the entity that has the mind and has the body is the 'I'. But now what exactly is this 'I'. Is it a self? And what is the human being? Is it a self attached to a mind and a body. Or is the self the mind? But if the self *is* the mind, how can we speak of its *having a mind?* I'm afraid we are confused.

Frank: [*chuckles*] Y'know, when the Lone Ranger and his Indian side-kick Tonto get captured by some Apaches, Lone Ranger says to Tonto, 'We're in real trouble', and Tonto replies, 'Who's this we, white man?' [*Descartes looks puzzled*] … Well, I'm not so sure as my friends that *we're* confused. I just think that *they're* confused. I think that the mind *is* the brain, and that the activities of the mind just are the activities of the brain.[2]

Look, sir, it was you who taught us that we can explain everything about living bodies by reference to the same general principles that govern physics – that the sciences of life are no different in principle from the physical sciences. Life, and the functions of living things, can be explained by reference to broadly speaking mechanical principles. That was a great insight. It freed neuroscience – the study of the brain and nervous system – from futile investigations of psychic pneuma, and from the ancient ventricular doctrine that located the physio-logical root of psychological functions in the ventricles of the brain.

Descartes: I am grateful to you for your compliments, Monsieur Frank. I agree with you that it was indeed an achievement of some moment. Now, *mes amis*, if you think carefully and methodically about your questions, it is not too difficult to discover the truth of the matter. It should be evident that you are not your body. For it is possible to doubt whether your body exists, but you cannot doubt whether you exist. And since that is so, you cannot be your body. If you were your body, then the fact that you cannot doubt that you exist would also mean that you cannot doubt that your body exists. But you can doubt whether your body exists.

Frank: But what about the brain. I can't doubt that I have a brain – I have a brain, and so does every other animal.

Descartes: *Non, non, mon ami.* You, who know that you exist merely by trying to doubt whether you exist, are not something you have. And can you really not doubt whether you have a brain? After all, you have never even seen or touched your brain. Of course, you have one. But that is hardly the soundest and most certain of knowledge, *n'est-ce pas?* You can, provisionally, in the course of your search for truth, doubt whether you have a brain, just as you can doubt whether you have a body.

Frank: [*a bit puzzled and out of his depth*] Well, I'm not sure how to answer you. I'm not a philosopher; just a scientist. But do go on, sir.

Descartes: *Bien!* The first step towards the true knowledge – *scientia* – that is absolutely certain is to say to yourself 'I doubt, that is to say: I think, therefore I am.' This establishes a truth indubitable concerning something that exists: *I* exist. Now ask yourself what is this 'I' by which you are what you are. It is a thing that thinks, is it not?

Jill: But I not only think, I affirm and deny, I want and intend, and many other things too.

Descartes: *Mais certainement!* But these are all modes of thought. By 'thought', I understand everything of which we are conscious as happening within us, in so far as we are conscious of it. So thinking is not merely reflecting, but also understanding, willing, imagining as well as sensory *experience* in general – that is, the *experience* of seeing, hearing and so forth, irrespective of whether one is dreaming, hallucinating or actually perceiving. All these I deem forms of consciousness.

Jill: So you are claiming that thinking belongs to our essence?

Descartes: *Exactement!* You are a mind or soul, the essence or nature of which is to think – to be conscious of what passes within you. The mind is a substance the essence of which is to think, just as body is a substance the essence of which is to be extended.

Jill: But even if it is true that thinking is the essence of the mind, it does not follow that *only* thinking is the essence of myself.

Descartes: I admit that what you say is true, Madame. But since I have a clear and distinct idea of myself in so far as I am simply a thing non-extended and thinking, and, on the other hand, I have a clear and distinct idea of body as a thing extended and unthinking, I know with certainty that I am distinct from my body.

Richard: So you are essentially a thinking substance?

Descartes: *Mais oui.* I am not a thought or collection of thoughts, as my friend David 'ume supposes. Thoughts demand a substance in which to inhere. One cannot have thoughts floating around like so many phantasms flitting through the air! I am a thinking substance.

Jill: But Hume argues that when, as he put it, he 'entered most intimately' into what he called *himself,* he always stumbled upon some particular perception or other – roughly speaking, what you call 'thoughts' – but could never perceive *a self,* a mental substance that persisted through time.

Descartes: *Le pauvre David! Certainement* he could not. The self is not something we perceive or experience. It is what thinks and perceives, imagines and wills. That thoughts inhere in thinking substances is not a matter of fact that we discover by experiences; it is presupposed by all thought and experiences. It is something that we clearly and distinctly perceive to be true by the natural light of reason.

Richard: So the self, in your view, is the same as the mind or soul and indeed the same as the 'I'.

Descartes: What you say is very true. And my mind is entirely distinct from my body, even though it is very closely intermingled with it.

[Richard tops up all the glasses, and thoughtfully takes a sip himself]

Richard: Well, I have some qualms. In the first place, you have simply bypassed the question we raised earlier, namely: how can I be something I have. I have a mind, of course, but by the same token I can't be the mind I have.

Descartes: *Mais* Richard, that is merely an idiom trivial. It can have no bearing on deep metaphysical questions.

Richard: Really? I'm not so sure … But let's pass over that one. I *suppose* I can't doubt whether I exist – although, I confess, I am not sure what the form of words 'I doubt whether I exist' mean. And for the sake of argument let me grant you that we can doubt whether our bodies exist, although I have qualms about that too – I mean, I don't even know how I'd go about doubting that I have a body. But even if I grant you all that, does it really show that I am distinct from my body? I mean, if a child knows what a triangle is, he cannot doubt that it has three angles, but he can doubt whether the sum of its angles equals two right angles. But *we* know that it is necessarily so. The child still lacks an adequate understanding of the nature and essence of triangles. Perhaps we lack an adequate understanding of mind and body and do not grasp that they are necessarily united and cannot exist one without the other.

Descartes: But, *mon ami*, you are no child, and you do know what the nature of matter is, namely a substance the whole essence of which is to be extended. And you know what the nature of mind is, namely: it is a thinking substance. Reflect moreover that the body, being an extended thing, is divisible – but the mind is utterly indivisible. When I consider myself in so far as I am merely a thinking thing, I am unable to distinguish parts within myself. I understand myself to be something quite single and complete.

Richard: Are you? Since you hold that you are your mind, surely your mind is divisible into different faculties – faculties of sense, of passion, of understanding and reason, of memory and so forth. And indeed you can lose one of these faculties, as when you suffer from amnesia. When you have lost your memory, you have lost a part of your mind.

Descartes: I think that I can get round this objection easily. We must distinguish between the faculty and what possesses the faculty. It is one and the same thinking substance that reasons and reflects, that senses and remembers. If you lose your memory, it is still one and the same thinking substance that before had the faculty of memory and now has lost it. Loss of a faculty does not imply divisibility of the thinking substance that previously possessed it. This one argument concerning the divisibility of the body and indivisibility of the mind would be enough to show me that the mind is completely different from the body.[3]

Jill: But surely the mind and the body are united in each human being? I'm not just a non-spatial mind temporarily embodied in a female human body.

Descartes: *Mais naturellement,* your body is not merely the vehicle for your soul. You are united with your body. You are, so to say, *intermingled* with it. Embodiment is not like being a sailor in a ship, who has to look to see whether the vessel in which he dwells is damaged. But you *feel* your sensations just *as if* they were located in your body.

Jill: But pains, itches and tickles *are* located in our bodies. We have headaches, stomachaches and backaches.

Descartes: *Non, non,* Madame Jill, you feel them *as if* they were in the body. But careful reflection on phantom limb pain shows that pain in the hand is felt by the soul not because it is present in the hand but because it is present in the brain.[4] We feel pain *as it were* in the hand or foot, but that does not show that pain exists outside the mind in the hand or foot.[5]

Frank: Yeah, that's right. Common sense tells us that a pain in the foot is really there in the physical space of the foot. But, just as you said, sir, we now know that's wrong. The brain forms a body image, and pains, like all bodily sensations, are parts of the body image. The pain-in-the-foot, as John Searle has written, is literally in the physical space in the brain.[6] But the way it appears to consciousness is just as if it were in the foot.

Descartes: What you say is very true, *Monsieur*. This judgement was already made by the author of *Principia Philosophiae*.[7] I am delighted that truth has prevailed.

Richard: I'm not convinced that what has prevailed in neuroscientific and in some philosophical circles too is the truth. The fact that an amputee can hallucinate that the pain he feels is in his foot when he has no foot does not show that when one feels a pain in one's foot when one *does* have a foot, the pain is not in the foot.

Descartes: But the pain is not located in the foot at all, *mon ami*. It is not in physical space as the foot is. And if you examine the foot you will not find a pain in it. The pain, in one sense, is in the brain, and in another sense, it is in the mind.

Frank: Yeah, that's dead right.

Richard: Well, I am afraid I disagree with both of you. It is the foot that hurts, not the mind. As Aristotle taught us, and as our language shows us, we have a sensitive body – our head may ache, our back may tickle, our leg may itch and our feet may hurt. The human body is not an insensate machine, but a living organism.

Descartes: I am very grateful to you, Monsieur Richard, for the objection. But as Monsieur Frank observed, the author of *Traité de l'Homme* showed quite clearly that the living organisms they are no more than machines.[8] And if you reflect further, you will note that our head does not have the headache, rather we do; our back does not have the backache, rather we have the ache that sensibly seems to us to be in the back; and so too, we may feel the pain that seems to us to be in the tooth, but the pain is caused by the decaying tissue in the tooth, which sends signals to the brain, which presents them to the mind in the form of the toothache.

Jill: You mean that we have toothache in the brain? That seems a very odd thing to say, Monsieur Descartes. When I have a toothache, it is my tooth that aches, not my mind.

Descartes: [*condescendingly*] Well, Madame, you may speak with the vulgar, but you should think with the learned. Sensation is a confused mode of awareness which arises from the union and as it were the intermingling of the mind with the body.[9] Furthermore, hunger and thirst, pleasure as well as pain, perception as well as sensation, like-wise arise from the union of mind and body. For this 'I', whereby I am what I am, is from the body entirely distinct. Nevertheless, a human being is an *embodied* soul or mind. It is not the accidental union of the two, but a substantial union. For a human being is an *ens per se*, in which mind and body form a unity. But sensation is not a part of my essence, that is, of the essence of my mind.[10]

Jill: [*indignantly*] And how is this 'intermingling', as you call it, effected, Monsieur Descartes? How can an immaterial non-spatial substance interact with, let alone intermingle with, a material substance that constitutes our body? You say that acts of will cause the transmission of animal spirits to the muscles, so making them contract or extend. And you say that the impact of light waves and sound waves on our nerves causes the movement of animal spirits from the sense organs to the brain, where they cause perceptual thoughts. But you don't explain how this can be.

Descartes: Madame Jill, you should read *Principia Philosophiae, la Dioptrique* and *Les Passions de l'Ame*. The author there explains that it is the soul that sees, not the eye; and it does not see directly, but only by means of the brain. Furthermore, the seat of the soul or mind – the point where it affects and is affected by the body – is not the whole brain, but the pineal gland that lies within the ventricles between the two hemispheres of the brain. Insofar as we do not see double or hear everything twice, there must necessarily be some place where the two images coming from the two eyes or the two ears can come together in a single image or impression before reaching the soul, so that they do not present to it two objects rather than one.[11] That place is the pineal gland.

Frank: Well, sir, that is not persuasive, y'know. First of all, animals other than us have a pineal gland too. But on your view they do not have the thought that things sensibly seem to them to be this or that, because they don't have any thoughts at all. Moreover, the pineal gland isn't located in the passage between the anterior and posterior ventricles and so can't affect the flow of 'animal spirits' between the ventricles. So your metaphor of the mind's being like the water engineer who turns the taps that control the flow of water to the fancy fountains at the Royal gardens is inappropriate. The pineal gland isn't in the ventricular fluid, and neuro-transmission isn't by means of animal spirits, as you supposed – it's electro-chemical. Secondly, your identification of the functional localization of what you call 'the soul' or 'the mind' is, *as a matter of scientific fact*, mistaken. It's the cortex that is the varied locus of mental functions, not the pineal body. We now know that different psychological functions have different cortical localizations. Vision, for example, has its functional localization in the 'visual' striate cortex, whereas thinking is associated with the prefrontal cortices.

Jill: Moreover, sir, your reasoning is flawed. Even if it were the mind that sees, then unless it sees the alleged image on the pineal gland, it does not matter whether there is one image there or two – or none at all. For whatever is there, if anything, *is not an object of vision*. It is not as if two images would cause double vision! As you yourself pointed out, there is not another pair of eyes within the brain[12] with which to see the representations you hold to be on the pineal gland – or, for that matter, on the visual striate cortex. Moreover, it is neither the brain nor the mind that sees and hears, smells and tastes. It is the creature as a whole, no matter whether man or animal. And animals really are conscious creatures, and they perceive things and feel sensations just as we do.

Richard: Whoa! Slow down. We needn't confront the matter of animal thinking and perceiving now. The central *philosophical* point we are advancing in criticism of your system, Monsieur Descartes, is that you have not given any explanation of *how* thinking immaterial substance can interact causally with unthinking and extended material substance. You insist, unlike your successors Malebranche and Geulincx, that they *do* interact. But you have not explained how that is possible – saying that the soul interacts with the body via the pineal gland simply shifts the problem back another stage, for now you must explain how the unextended immaterial soul can interact causally with the extended material pineal gland, or, as we should say, with the cortex.

Descartes: Ah, *je suis mortifié*. Her Highness, the Princess Elizabeth of Bohemia, once sapiently asked me how the soul, being only a thinking substance, can determine the animal spirits to bring about voluntary actions. I replied to her that the question she posed is the one which can most properly be put to me in view of my published writings.[13] Everyone invariably experiences in himself, without philosophizing, the union of the soul and the body. I concede to you that the human mind is not capable of forming a very distinct conception of the distinction between the soul and the body and also of their union. For to do this it is necessary to conceive them as a single thing and at the same time as two things. This is an *absurdité*. But everyone *feels* that he is a single person with both the body and the thought so related that the thinking mind can move the body, and the mind can also feel the things that happen to the body.[14]

Frank: Well, Monsieur Descartes, now y' really have blown it. If reasoning comes up with one thing and feeling with the opposite, then one may be right and the other wrong, or both may be wrong, but they can't both be right.

Jill: Quite apart from that, I don't think that your appeal to *feeling* the unity of mind and body is warranted. We don't feel that the thinking mind can move the body – we know that *we* can move our limbs, and indeed move ourselves. That is not a form of telekinesis! My thinking mind can no more move my arm than it can move my armchair. It is *I* who move my arm, not my mind.

Richard: Moreover, the issue is not so much *whether* our mind and our bodies interact. We are not asking you to persuade us of that, sir. What you owe us is an explanation of *how* they can interact.

Descartes: I am not in the habit of crying when people are treating my wounds, and those who are kind enough to instruct me and inform me will always find me very docile. I cannot answer your question, *Messieurs*. But before I take my leave, I wish to remind you of one further point. We are all here *in Elysium* – so the soul and the body *must be* separable, for our mortal remains have long since turned to bone and dust. But our souls, being indivisible, are therefore also indestructible, and hence immortal – unless, of course, reduced to nothingness by God. So I must surely be right, even though it is beyond my powers to explain how it is that I am right. Perhaps, as my follower Noam Chomsky says when he encounters what he calls 'Descartes's problem' of how we can utter sentences we have never encountered before, it is beyond the powers of the human mind to comprehend such matters.[15] [*He rises to his feet, picks up his hat, and bows to the others*]

Au revoir, Madame Jill, Au revoir, Messieurs. [He leaves, rather miffed]

Frank: Well, that was kind of interesting, but I don't think it helped much. The idea that the mind is an immaterial substance interacting with the brain is just hopeless. But he did make a point that floored me. What d'you make of the idea that the mind is immortal since it is indivisible, Richard?

Richard: Well, assuming that all destruction is disintegration or decomposition, then if the mind were indivisible, it would be indestructible. But although one might say that the mind is not divisible, one must admit that it is not indivisible either.

Frank: What d'you mean?

Richard: Well, it is not as if you cannot divide a mind no matter how hard you try. It's rather that there is no such thing as dividing a mind, just as there is no such thing as dividing red. Minds, like colours, are neither divisible nor indivisible. So I wouldn't hang any hopes of immortality on that argument, old chap.

Frank: I see. Yeah. I always thought that dualism is a false theory. Y'know physics has disproved it.

Richard: [*incredulously*] Say that again!

Frank: Well, if my mind could move my body, that would violate the law of conservation of momentum.[16]

Richard: No, no, Frank. That objection was already advanced by Leibniz, but it is wholly misconceived. No scientific discovery and no scientific theory can resolve a philosophical or conceptual problem.

Frank: [*annoyed*] What the hell do you mean?

Richard: There's no need to get hot under the collar, old boy. You grant, I trust, that no discovery in physics and no law of physics can contribute to the resolution of a problem in pure mathematics, let alone confirm or infirm a mathematical proof. After all, it's not as if Newton's physics confirmed the theorems of the differential calculus.

Frank: [*still resentful*] OK, sure. So what?

Richard: Well, that's because mathematics is concept-formation by means of proofs. And philosophy is, among other things, concept-clarification by means of linguistic description. Conceptual clarification is not answerable to facts and theories of physics. Concepts, including the concepts of physics, create the logical space within which physics can determine facts and formulate theories concerning matters of fact.

Frank: OK. But even if I give you that, it still seems to me that if Descartes were right, then the mind's making the body move by acts of will *would* violate the laws of physics, in particular the law of conservation of momentum.

Richard: No, *if* the mind could move matter, it would show definitively that the law of conservation of momentum is false. But this hypothesis can only be true or false if the statement that the mind moves the body by acts of will makes sense. And that would make sense only if the idea of an immaterial substance made sense, and if we could render intelligible the supposition that an immaterial thing can have causal powers and could bring about change by acting *on* a material thing.

Jill: You mean the idea of an immaterial mind makes no sense?

Richard: Just so. There is no such thing as individuating substances independently of their being material space-occupants. Abstract material constitution, and with it spatial location and a spatio-temporal path through the world, and we would have no principle of individuation or criteria of identity.

Jill: I don't follow.

Richard: The deep trouble with Cartesian minds is that we cannot distinguish between one mind having a thought, and a thousand different minds having the very same thought. And, as Kant pointed out, we cannot distinguish between one mind persisting over time and having fresh thoughts from a thousand successive minds, each with the same thoughts as its predecessor with the addition of one more thought. So every time one has a thought, a different mental substance springs into being, possessed of all the thoughts of the prior substance together with a new thought – just like one billiard ball passing its momentum on to another on impact. To put the matter technically, Cartesian minds lack criteria of synchronic and of diachronic identity but if we don't know what counts as one mind and what counts as a thousand different minds, then we don't know what we are talking about.

Jill: But that's absurd!

Richard: Of course it is. But that's no thanks to Descartes's tale, since on his account it cannot be excluded. The very notion of an immaterial mental substance makes no sense.

Frank: That's OK by me! I told you that the answer is clear. Dualism in any shape or form is just plain wrong. We have to opt for straightforward materialism: *the mind is the brain.* I have a brain, and *that is what it is* to have a mind. Because all the mental functions of perceiving, thinking, imagining, deciding and willing are brain functions. Descartes was right to think that the body is a machine. He was only wrong to think that the mind is an immaterial substance that interacts with it.

Richard: [*laughing*] You mean it's not elephants all the way down, as Russell's old lady thought, but it's machinery all the way up![17]

Frank: [*chuckles*] Yeah, that's right. The fact of the matter is that although we feel ourselves to be in control of our actions, that feeling is the product of our brain, whose machinery has been designed, on the basis of its functional utility, by means of natural selection. We are machines, but machines so wonderfully sophisticated that no one should count it an insult to be called such a machine.[18]

Jill: Oh come on, Frank. You can't really believe that!

Frank: Why not? I do believe it.

Jill: Well, because machines are neither conscious nor unconscious. They take no pleasure in what they do and suffer no pain. They neither love nor hate. They do not deliberate on courses of action and then decide what to do on the basis of reasons. Machines don't know the difference between right and wrong, they have no obligations and they have no rights either. It's no insult to call a machine a machine, but it certainly is an insult to call me a machine! Do you think that I have no moral sense? Or are you suggesting that I have no rights and duties?

Frank: No, no. Of course not.

Jill: Moreover, the fact that, in one sense, everything we think, feel and do depends upon the activity of our brain, as you say, does not show that it's the brain that thinks and feels, makes decisions and acts.

Frank: Why not? We perceive because the brain forms an internal image on the basis of the information it receives from the senses. That's scientific fact.[19] All y' have to do is look it up in any decent textbook of neuroscience. We know that the brain makes decisions before you're even conscious of it. Benjamin Libet showed that decades ago.[20] And it's the brain that makes our hands and legs move and do things. In fact, we are our brains. As Chris Frith says, we are nothing more than 1.5 kilograms of sentient meat that is our brain.[21]

Richard: No. That isn't scientific fact, it's scientific confusion. For heaven's sake, Frank, you weigh more than one and a half kilos, and you're taller than seven inches. To be sure, you have a brain, but you're not what you have, and your brain doesn't have a brain – it is one. Your brain is in your skull, but *you* are not enskulled. You can't seriously believe such nonsense. Moreover, there is no such thing as a brain's forming images on the basis of information it receives from the senses. When we observe the world around us, what we see are objects and their properties, the unfolding of events and the obtaining of states of affairs. We don't see images, unless we are in a picture gallery. Nor does the brain *receive information*, in the sense in which you or I might receive information about lectures and concerts here by reading the *Elysian Gazette* or the *Heavenly Herald*.

Frank: Y'mean that the theories of Nobel Prize winners like Eric Kandel, Frank Crick, Gerald Edelman, as well as the theories of world-renowned scientists like Michael Gazzaniga, Antonio Damasio and Horace Barlow, are plain false? I mean, what scientific work have you ever done, Richard?

Richard: None at all, my dear fellow. But I am not saying that their theories are false.

Frank: [*irritated*] You're just advancing your opinion against theirs. There is a whole group of representationalists in cognitive science and the philosophy of cognitive science who think it is perfectly OK to attribute – for example – memory to cognitive subsystems. It is, to put it mildly, provocative to suggest your view as if it were a settled fact that they are wrong. That's just opinion presented as fact.[22]

Richard: My dear chap, it is no more an opinion than it is an opinion that red is a colour or that nothing can be both round and square at the same time. Is it your opinion that red is a colour? Come now! Nor am I saying that it is a matter of fact that brains don't think or remember or form images. If something is a matter of fact, then things *are* so, but they might have been otherwise. But this is a matter of logic, not of fact. It makes no sense to say that the brain thinks or remembers. And that is why these theories are not false. They are nonsense.

Frank: [*spluttering with indignation*] Who the hell are *you*, Richard, to say that the work of these distinguished scientists is all rubbish? That's just outrageous.

Richard: Calm down, Frank. I didn't say their work was rubbish. I said their claims about the brain that we just mentioned are nonsense. It makes *no sense* to say that the brain thinks, remembers, and perceives, constructs hypotheses and guesses what is, as your friends misguidedly put it, 'out there'. It is senseless to say that the brain decides and wills. These are forms of words that look as if they make sense, but don't. They are concealed forms of nonsense.

Frank: I don't see why. That's the way neuroscientists talk. What's wrong with it? It may be excluded from folk-psychological language, but it sure isn't excluded from scientific language. And who the hell are you to tell scientists how they should talk?

Richard: Frank, they may talk as they please. If they want to talk nonsense, let them talk nonsense. I'm not a linguistic policeman. All I am pointing out is that if you chaps want to speak *of yourselves* as thinking and reasoning, perceiving and feeling, deciding and acting in the received sense of these terms, then you cannot coherently *also* say that *your brain* thinks and reasons, perceives and feels, and so forth, *in the same sense.*

Frank: I don't follow you.

Richard: It is actually straightforward, once you orient yourself correctly, Frank. Look, you don't say that the table feels things, do you.

Frank: No, I'm not stupid, y'know.

Richard: No, of course you're not. You're one of the most intelligent scientists I know, old chap. Now, you don't think that the trees and roses over there can see or hear, do you.

Frank: [*a bit mollified*] No, of course not.

Richard: Why not?

Frank: Why not? Well, for one thing, they don't have eyes.

Richard: Quite so. And they don't duck when you throw a rock at them. Nor do they look at things or move closer to observe things better.

Frank: Yeah. OK. So?

Richard: Well, does the brain have eyes? Does it look at things that catch its interest? Does it move its eyes to follow what it's looking at? Does it rub its eyes when there's a glare, and shield them when it's too bright?

Frank: No. If it has no eyes, it can't engage in visual behaviors.

Richard: That's right. There is no such thing as *a brain* exhibiting visual behaviour. Increased neural activity in the 'visual' striate cortex is not a form of behaviour. Moreover, even if, as you suggest, brains see and hear, think and remember, how would that help *you* to see and hear, think and remember?

Frank: Well, my brain informs me.

Jill: But, Frank, how can your brain inform you of anything if it cannot speak English? Or are brains language-users? Do brains have voices? And how would you listen to your brain when it talks to you? With an inner ear? It is schizophrenics that hear inner voices, you know.

Frank: Yeah, OK. I see what you're driving at. So, how does all this add up?

Richard: How it adds up is this: It only makes sense to ascribe vision to beings that have eyes with which to see, and that exhibit – manifest – their visual powers in their behaviour. An animal that sees avoids obstacles in its pathway; it goes around them or steps over them. It examines and scrutinizes anything it sees that rouses its curiosity, and it flees from dangerous things it perceives. It's not false that trees see. If it were false, then it might have been true but doesn't happen to be. But trees aren't blind either. Nor are brains – they can neither see, *nor are they blind*. It *makes no sense* to ascribe seeing or overseeing or not seeing to them. Brains don't make decisions, and they're not indecisive either. It simply lacks sense to say that my brain decided to do something. And the brain doesn't *make* your hands and legs move, rather it makes it possible for you to move them at will.

Frank: So what you're saying is that cognitive neuroscientists are making a conceptual mistake here.

Richard: Exactly. That's why I said that they're talking nonsense. I don't mean that they're talking sheer rubbish; what I mean is that they're transgressing the bounds of sense and that they're putting words together in a way that's excluded from the language. They're committing a mereological fallacy.

Jill: What's a mereological fallacy, Richard?

Richard: Mereology is the logic of parts and wholes. One kind of mereological fallacy is ascribing to a part of a thing properties that can only intelligibly be ascribed to the thing as a whole. Look, an aeroplane can't fly without its engines, but it isn't its engines that fly, it's the 'plane. An antique bracket clock can't keep time without a fusée, but it's the clock that keeps time, not the fusée. So too, an animal cannot walk or run, talk or sing, without a brain, but it's not the brain that walks and talks, it is the animal as a whole.

[*Silence for a moment*]

Frank: OK. You're a clever beggar, Richard. I see what y'mean.

Richard: Good! Well, if you see that, old boy, then you should also see why neuroscientific materialism is a degenerate form of Cartesian dualism. It just replaces ethereal minds with grey glutinous stuff and leaves everything else intact. It replaces mind/body dualism with brain/body dualism.

Frank: Hey, wait a minute. The brain is just as material as the rest of the human body. There can't be any such thing as brain/body dualism. Both the brain and the body are material.

Richard: Of course. But you neuroscientists leave intact the whole structure of Cartesian dualism. You think that the only thing wrong with it is that it introduces a mental substance as the subject of all psychological attributes. You think that perceiving is the generation of mental images in the brain by the impact of material things and of sound or light waves on our sense organs. That is senseless. You think that voluntary movement is movement caused by the brain's deciding to move. And that is nonsense too. And you imagine that thinking is information processing by the brain, whereas it is nothing of the sort. Frank, the trouble about you neuroscientists is not that you are anti-Cartesian, it is that you are not nearly anti-Cartesian enough.

Frank: OK. I get the message. But where does that leave us?

Jill: It means that there aren't enough 'or-s' in your battle cry: Either dualism or materialism.

Frank: OK. So what's the new deal?

Jill: Yes. What is the third way, Richard? We haven't even scratched at answers to the questions we started out with. I mean: What is the mind? What is the self? What is the Soul? What am I, and who or what is it that has a mind, a self or a soul? And how is the mind related to the body? And who is it that has a body? And …

Frank: Enough already. You're swamping us. We need some help.

Richard: Yes, a *deus ex machina* would come in useful.

[*A dignified, good-looking, late-middle-aged gentleman strolls in from the trees, wearing an ancient Greek himation and sandals. He has a well-cut beard, flecked with grey, and greying hair. They all recognize Aristotle, and the men rise to their feet*]

Aristotle: Good day. May I join you for a short while. I think I may be able to help you a little.

Richard: We are honoured that you should choose to join our modest symposium here, sir. May I introduce my friends? This is Jill; this is Frank and my name is Richard. May I offer you some wine? [*He pours Aristotle a glass of wine and hands it to him*] It comes from an excellent cellar: the *Nectarian*. Please sit down. [*Aristotle takes a seat*] We have been struggling to clarify the nature of the mind.

Aristotle: Your difficulties are quite reasonable. For among the many, and even more among the wise – including Plato – there is division of opinion and obscurity of statement concerning the mind.

Jill: Well, we certainly have plenty of difficulties. We had a long discussion with Monsieur Descartes, and examined his view that the mind is a separate substance from the body and that its essential nature is thought, construed as consciousness of what passes within us. But there were many objections to this dualist view. And we examined Frank's view that the mind is just the brain and that all the attributes of the mind are in effect attributes of the brain. But Richard showed us that this view too is unacceptable. We really are at a loss.

Aristotle: But that is already progress, madam. It is the height of madness not merely to be ignorant but not to realize that you are ignorant, and therefore to assent to false conceptions and to suppose that true conceptions are false.

Jill: But there are so many competing reasons that it is difficult to know where to begin.

Aristotle: I agree with you, madam. But you must bethink yourself that some people offer reasons that are irrelevant or unsound, and often get away with it. Some people do this in error. Others are sheer charlatans. By such arguments even thoughtful people may be caught out by those who are lacking in the capacity for serious theoretical reflection.

Richard: So where do we begin?

Aristotle: Most of the controversies and difficulties will become clear if we offer an appropriate explanation of how to think of living beings. Living beings are organisms. Where we have living beings, beings that may prosper and flourish or deteriorate and die, we have welfare and ill-fare. Where we have organisms of developed form, we have organs. Where we have organs, we have function and purpose. Where we have function and purpose, welfare and ill-fare, we have varieties of the good. For we have the goodness of health, the goodness of organs, the goodness of their exercise, the good of the being that has organs and the goodness of that which is conducive to the good of the being. From this it is evident that one of the roots of axiology is biology.

Jill: So you mean that the sciences of life are inseparable from the study of the good?

Aristotle: But of course. No one in his right mind could think otherwise. However, let us focus upon our task, which is to clarify what is distinctive of all living beings.

Frank: Well, I guess it's that they all ingest nutriment from their environment, they all grow and reproduce, giving rise to the next generation.

Aristotle: I agree. We must begin our investigation by noting the *archē*, the distinctive principle, of the lower forms of life – the plants. It is evident that they have the powers of metabolism, growth and reproduction. This is characteristic of all species of living things, is it not? These nutritive or vegetative powers constitute the form of botanical life. Indeed, we may say that they *inform* the organism, constituting the essential powers of botanical organisms that have organs. And we may characterize this form as the nutritive or vegetative *psuchē*.

Richard: But that is not how Plato thought of the *psuchē*? He thought that the soul is something that resides temporarily in the body and that will leave the body on death.

Aristotle: Indeed. Like many others, Plato joined the *psuchē* to a body, or placed it in a body, without explaining the reason for their union or the bodily conditions required for it. He thought that the *psuchē* is *embodied*. But that is absurd.[23] For it is not as if *any psuchē* could be conjoined with *any* body – the *psuchē* of a man with the body of a tree, for example, or with the body of a bean – as the Pythagoreans supposed. This is not a helpful way of thinking about the soul. We should not conceive of the *psuchē* as a being – a secondary substance of a strange kind – but rather as the form of living things. The *psuchē* is not embodied, rather the organic body – the body with organs – is *empsuchos*, ensouled. The *psuchē* is constituted by the distinctive powers that *inform* living beings and in virtue of which they are the kinds of beings they are. Thinking thus, we have a far more powerful way of conceiving of natural life in general and of ourselves as part of nature – albeit partaking of the divine or blessed.

Frank: Hey, slow down. What do you mean 'partaking of the divine'?

Aristotle: Mankind possesses nothing divine or blessed that is of any account except what there is in us of mind and understanding. We are born for two things, understanding and action, and we fully realize our nature in the exercise of our understanding in the noble endeavour to comprehend the world in which we pass our lives, on the one hand, and in the excellence of our actions in accordance with virtue. To achieve this to the best of our abilities is what I mean by 'partaking of the divine'.

Frank: I see. So you don't mean that the mind or soul, or the *psuchē* as you call it, is *a part* of a living animal? If it was a part, it might be separated from the body – and that's surely just a fiction.

Aristotle: The *psuchē* is the form of living things. Where there is no living thing, there can be no *psuchē*, for the being of such forms is to inform matter. Now, we can say that a substance has parts in many different senses. It is clear that the *psuchē* is not a part of a body that potentially has life as wheels are a part of a chariot. The *psuchē* stands to the organism somewhat as the shape of a statue stands to the marble of which the statue is carved. That is why it is absurd to ask whether the body and soul are one or two. That is like asking whether the wax and its shape are one or two.[24] From this it is clear that the *psuchē* is inseparable from the body.[25] It is the principle of life characteristic of kinds of living beings, for its distinctive powers are what make a living being with organs the kind of being it is.

Jill: So according to you, sir, the *psuchē* explains the nature of life?

Aristotle: It is a notion that belongs to the sciences of life, but not after the manner of those who conceive of the soul as corporeal and originative of movement, and identify it with hot breath or hot blood, thinking of these as the principle of life. The *psuchē* characterizes organic life by reference to its powers. Of course, we can study the powers of living things only by studying their behaviour, for activities and actions are logically prior to potentialities. But let us not jump ahead of ourselves as a steed jumps before it has reached the correct distance from a wall. We must proceed methodically, in the correct order. So, the nature of the nutritive *psuchē* by which all living things are informed has been outlined. Now, what further powers characterize *animals*?

Frank: Well, I guess they can perceive their environment, they have desires and they can move about to get what they want.

Aristotle: Quite so. Animal life, over and above the powers of the nutritive *psuchē*, is characterized by sensibility, of which the primary form is touch and hence taste. Where there is sense, there is the capacity for pleasure and pain. Where these are present, there too must be appetite. Otherwise even the most primitive of sea animals could not nourish themselves and distinguish what is beneficial from what is detrimental to them. Certain kinds of animal also possess powers of locomotion, and further senses of sight, hearing and smell. Where there is sensibility and self-movement, there too must be desire and aversion, and action for the sake of a goal.

Frank: Y'mean that animals have two of these *psuchē*-s – two souls. That seems weird!

Aristotle: Not weird, absurd. What I mean is that the distinctive set of powers of animate creatures *includes* the essential powers of vegetal forms of life, namely: powers of nutrition, growth and reproduction, but incorporates further distinctive powers that constitute, in one sense of the word, the essence of animal life. And, of course, we classify different kinds of animals according to their distinctive powers and the distinctive organs by means of which they exercise them.

Jill: But still, there is something distinctive of us humans in virtue of which we conceive of ourselves as having a mind.

Aristotle: Of course. We possess rational faculties. Animals lack the powers of reason, calculation and reflection. And it must be born in mind that thought is found only where there is also reason. What distinguishes us within the realm of nature is the possession of a rational *psuchē*, over and above the nutritive and sensitive *psuchē*. It is this that you may think of as *the mind*, which is distinctive of mankind. The power of reason is the ability to apprehend the transition from premises to the conclusion that they determine, and hence too the power of understanding the manifold 'because-s' that answer the question 'Why?'. Only beings that can answer the question 'Why?' can be answerable for their deeds, and know the difference between virtue and vice. Rationality is exhibited in drawing inferences from premises and in deriving conclusions from evidence. It is manifest in deliberating, in rational choice and in sensitivity to reasons.

Richard: I don't follow that. I'm not sure what you mean by 'sensitivity to reasons'. Since sensitivity is itself a potentiality, not an actuality, I am not sure what you mean by saying that mind is exhibited in a potentiality, given that you also want to say that it *is* a potentiality.

Aristotle: You must realize that there are many different kinds of 'can' and different sorts of potentiality. Because mankind is endowed with reason and understanding, we can understand something as warranting thought and action. We can apprehend a 'this is so' as a justification for acting thus-and-so, or as a warrant for concluding that things are so. But we may know something to be a reason, just as we know something to be so, even when we are asleep. Or we may apprehend something to be a reason while we are awake, and yet not take notice of it. Or we may notice it and act immediately without deliberation, as when we catch someone who is about to fall. Or we may apprehend something as a reason, and deliberate on what is to be done, and later do it for that reason. What is clear is that in all or some or one of these ways, we, unlike all other animals, are sensitive to reasons.

Frank: I don't see how this 'rational *psuchē*' or mind can interact with the body if it isn't a part of the body, like the brain is. I mean if the nature of the rational *psuchē* is to be sensitive to reasons, how does it make the body move?

Aristotle: We are speaking of powers, my dear sir, not of things. The ability of an axe to cut is not a part of the axe. Nor does it interact with the axe, or make the axe cut. We cannot see without eyes, but eyesight is not a part of the eye. It is not the eye that sees, it is the animal with eyes. Without eyes, there is no eyesight, but eyesight does not make the eyes see. So too, it is the human being that reasons and deliberates for reasons and acts on account of reasons and for the sake of rationally chosen ends. It is not the mind or rational *psuchē* that reasons, infers and comes to conclusions. It is the human being. To say that the *psuchē* reasons or deliberates is like saying that the *psuchē* weaves or builds. Surely it is better not to say that the *psuchē* pities, learns or thinks, but that the man does these things with his *psuchē*.[26]

Jill: You mean one does things with one's mind or rational *psuchē* just as one sees with one's eyes and walks with one's legs?

Aristotle: No, my dear lady, not at all. The *psuchē* is not a part of a living being and so the rational *psuchē* is not an organ of a human being like the legs and eyes. One does things with one's *psuchē* in the sense in which one does things with one's talents.

Jill: Ah, I see. So the very question of how the mind is related to the body is itself a misguided question?

Aristotle: Of course. It is akin to the question of how the potter is related to his ability to throw a pot, or of how the eye is related to eyesight. These are not relations at all. We have capacities and abilities, liabilities and susceptibilities, but while *having an axe* is a relation between an owner and his possession, having powers is no relation. You must think of the peculiarities of the idea of *having*. For we speak of *having* in a number of different ways: of having knowledge, which is akin to possessing abilities; or of having courage, which is a trait of character; of having a height or length, as well as of having a cloak or tunic, as when one covers oneself after exercise in the palaestra. And we speak of having a ring on one's finger, as when one wears a ring on a part of oneself. We speak of the jar as having wine in it, when it contains wine. And there is an even stranger way of having, as when we speak of having parents – which means that one's parents are alive, and of having a wife, which signifies no more than that one is married to her.[27]

Richard: Yes, I read that in your *Categories*. But you wrote there that you had made a pretty complete enumeration of the different ways 'having' is spoken of. But you didn't speak of having a mind, of having something in mind or having something at the back of one's mind. Nor did you extend your remarks to such forms of *having* as having a thought, having a reason or having a goal. But that surely is a most fruitful way of pursuing further your idea of conceiving of the mind, the *rational psuchē*, in terms of having first- and second-order abilities and exercising them. For here too we must examine carefully what lies behind all this *having*.

Aristotle: But of course. I wrote three carefully composed pages of notes in Chapter 15 of the *Categories*, which made the matter quite clear. You don't mean to say that you stopped before the end?

Richard: Well, no. I read right to the end, but Chapter 15 consists only of one brief paragraph.

Aristotle: [*agitated*] You mean that those concluding pages have been lost?

Richard: Well, yes, I suppose they must have been.

Aristotle: [*jumping to his feet*] This is grievous indeed. I did not realize that the complete notes have not survived. Pray excuse me now. I must see whether there is a decent copy in the Library here. [*He pauses and collects himself*] But before I go, let me suggest to you how to pursue matters further for yourselves. Bear in mind that if you begin with things that are said in a manner that is true but unenlightening, you will make progress towards enlightenment by constantly substituting more perspicuous expressions for the ones that are more familiar but confusing. To make clear distinctions is not characteristic of most men. But this is what you must at all times strive to do. Now I must go and try to find Theophrastus and Neleus to see whether they can throw some light on this loss. Farewell. [*He leaves*]

Jill: Oh what a shame! That was wonderful. And we were just about to get to the heart of the matter.

Frank: Who are these guys Theophrastus and Nellyus?

Richard: Neleus. Theophrastus took over the Lyceum after Aristotle died, and so he inherited all of his manuscripts, and he passed them on to his nephew Neleus of Skepsis, one of Aristotle's last pupils, for safe keeping.

Frank: Ah! ... Well, that was some display of fireworks. He sure did leave dualism in tatters.

Jill: But also reductive materialism, Frank. What he's offering us is naturalism without reduction – a conception of the mind that is neither dualist nor materialist.

Richard: Yes. What we now have to do is apply the schema he's given us to our normal discourse about the mental. We need to examine the use of such phrases as 'having a thought at the back of one's mind', 'having a thought cross one's mind', 'having something in mind', 'making up one's mind' and ...

Frank: Hey, wait a minute. If he's right, then it makes no sense to speak of the mind being separable from the body.

Richard: Yes, that seems eminently plausible to say the very least. Surely you must find that idea congenial?

Frank: Yeah, sure. But then how do you reply to Descartes's parting shot? I mean, we *are* here y' know, and our bodies must have turned to dust by now. So maybe we do survive without the bodies we once had.

Richard: My dear chap, did it never occur to you that we might simply be characters in someone's dream?

[*There is a roll of thunder and flash of lightening, and all goes black*]

Notes

1   F. Crick, *The Astonishing Hypothesis* (Touchstone, London, 1994), p. 3.
2   Ibid., p. 7.
3   Descartes, *Meditations* VI (CSM II, 59; AT VII, 86).

4   Descartes, *Principles of Philosophy* IV, §196.
5   Ibid., I, §67.
6   J. R. Searle, *The Rediscovery of Mind* (MIT Press, Cambridge, MA, 1992), p. 60.
7   Namely Descartes, in his correspondence he liked to refer to himself as 'the author of …'.
8   Descartes, *Treatise on Man* (CSM I, 108–9; AT XI 202).
9   Descartes, *Meditations* VI (CSM II, 56; AT VII, 81).
10 Ibid., 73.
11 Descartes, *Passions of the Soul* (CSM I, 340; AT X, 353).
12 Descartes, *Optics* (CSM I, 106; AT XI, 119).
13 Descartes, letter to Princess Elizabeth, 21 May 1643 (CSMK 217; ATR III, 663–64).
14 Descartes, letter to Princess Elizabeth, 28 June 1643 (CSMK 228; AT III, 694).
15 N. Chomsky, *Language and Problems of Knowledge* (MIT Press, Cambridge, Mass., 1988), pp. 147–52.
16 Paul Churchland, 'Cleansing Science', *Inquiry* 48 (2005), 464–77.
17 Russell relates that at one of his lectures he remarked on the futility of the Hindu supposition that the earth rests on the back of a great elephant. After the lecture, a little old lady approached him and said, 'You've got it all wrong, Lord Russell. It's elephants all the way down!'
18 Colin Blakemore, *The Mind Machine* (BBC Publications, London, 1988), pp. 270–2.
19 Antonio Damasio, *The Feeling of What Happens* (Heineman, London, 1999), p. 320; G. Edelman, *Bright Air, Brilliant Fire – On the Matter of the Mind* (Penguin, Harmondsworth, 1994), p. 119; E. R. Kandel and R. Wurtz, 'Constructing the Visual Image' in E. R. Kandel, J. H. Schwartz and T. M. Jessell (eds), *Principles of Neural Science and Behavior* (Appleton and Lange, Stamford, CT, 1995), p. 492.
20 B. Libet, *Neurophysiology of Consciousness* (Birkhäuser, Boston, 1993).
21 C. Frith, 'My Brain and I', *Nature* 499 (18 July 2013), p. 282.
22 Kim Sterelny, email to Harry Smit, 29 July 2013.
23 Aristotle, *De Anima* I, 407^{b}14–26.
24 Aristotle, *De Anima* II, 412^{b}5–9.
25 Ibid., 413^{a}4.
26 Aristotle *De Anima* I, 408^{b}12–15.
27 Aristotle, *Categories*, chap. 15, 15^{b}18–32.

SUPPLEMENTARY READING

Aristotle, *De Anima*.
Descartes, *Meditations* 2 and 6.
P. M. S. Hacker, *Human Nature: The Categorial Framework* (Blackwell, Oxford, 2007), chap. 8.
A. J. P. Kenny, *The Self – The Aquinas Lecture 1988* (Marquette University Press, Milwaukee, 1988).
A. J. P. Kenny, 'The Geography of the Mind', repr. in his *Essays on the Aristotelian Tradition* (Clarendon Press, Oxford, 2001), pp. 61–75.

Second Dialogue

THE MIND AND THE BODY

Protagonists:

Frank Craik: a contemporary American neuroscientist, casually dressed in jeans and pullover, with open-necked shirt. He has a marked American accent.

Jill: a philosopher in her early thirties, dressed in an elegant but informal manner.

Richard: a middle-aged Oxford don of the mid-twentieth-century, dressed in cavalry twill slacks, waistcoat and tie, and well-cut jacket.

Alan White*: a small man, dressed in a dark blue jacket and grey slacks, with a light blue shirt and a bow tie. He speaks with a slight Irish accent.

Peter Strawson: dressed in a well-worn tweed suit and a woollen tie. He has a rather deep voice, speaks slowly and very deliberately.

The setting is a garden in Elysium. The sun is shining. A rich verdant lawn is surrounded by flower beds and flowering bushes, with a grove of magnificent trees behind. Beyond, there is a large lake, and in the distance, high mountains. Five comfortable garden chairs are placed in the shade of some trees. There is a low table on which are placed a wine decanter and glasses, three of which are half full. Richard, Jill and Frank are deep in conversation.

Frank: OK. I agree that our discussion with Aristotle showed us why dualism and crude materialism are wrong. He made clear what the mind is, but he still left us with the problems we started with.

Jill: What problems do you have in mind, Frank?

Frank: Why, the problems that Richard raised: why do we speak of *having* a mind? And then there is the further question which we didn't

---

even have time to discuss with Aristotle: why do we speak of *having* a body?

Jill: Yes. We really must talk about that. I *have* a body, just as I *have* a mind. But what exactly is this relation of *having*? And what exactly is involved in having a mind and a body?

Richard: Well, at least this much became clear: Having a mind is to be compared with having powers and abilities, with having talents – not with having a house or a car. And doing something with one's mind is not like doing something with a hammer or a screwdriver, but like doing something with one's talents. It is exercising one's intellect and will. And what it is that has a mind is obviously the living human being.

Jill: All right. So far so good. Aristotle certainly made that clear. But what about *having a body*? If I *have* a body, how can I also *be* a body? Surely I can't *be* what I *have*? And surely *having a body* is not at all like having talents. Having a beautiful body isn't exercising anything, even though it is very attractive.

Frank: OK. I can see that. But I don't see what you're driving at?

Jill: The point is that to say that I have a body is to imply that I *own* my body. And if I own my body, then, to be sure, I have a right to my body, and a right to determine how I use it. Women have a right to their bodies.

Richard: Well, I'm not sure what that is supposed to mean. You're rushing ahead far too quickly, Jill. I think we should slow down and follow Aristotle's advice.

Jill: Namely?

Richard: Well, when I raised the issue of the idiom of 'having', as in 'having a mind' or in 'having a thought at the back of one's mind', he said that we should begin with things that are said in a manner that is true but unclear and replace the confusing expressions by ones that are more perspicuous.

Jill: All right. So how are we to deal with 'having a body' and 'my body' or 'her body'?

Richard: No, no. Take it slowly, Jill. We should start with idioms related to *having a mind*.

Jill: I don't see why? I'm much more interested in exploring what is involved in having a body, and what the moral consequences of being in possession of one's body are.

Richard: Jill, we'll come to that. Look, we have been given some pretty powerful insights into the nature of the mind. We should build on that. For doing so may also give us some clues about the nature of the body and of having a body. Even more importantly, it will make clear how to proceed. For we don't want to 'consult our intuitions'. We want to examine our concepts.

Jill: What is wrong with examining our intuitions? Most women have a very powerful intuition that they own their body, that it belongs to them.

Richard: We'll come to that, Jill. But we'll be on much safer ground if we first prove our methods and sharpen our techniques by examining idioms of possession – of 'having' – in relation to the mind. Intuitions are really neither here nor there. After all, what is an intuition anyway? It's either just a firm conviction, which may or may not be a strong prejudice, or it is a hunch or guess. And none of these has any place in philosophical inquiry. Pre-philosophical convictions are just grist for philosophical mills.

Jill: You have forgotten the most important form of intuition, Richard, which is direct insight into a truth. To be sure, is that not what female intuition is?

Richard: But Jill, that merely moves the question back a stage. For now you have to have a criterion to distinguish between genuine intuition and apparent intuition. To draw that distinction we need an independent determination of the correctness or truth of what has allegedly been intuited – which in our case is the proposition that we own our bodies. But that is where we started. So let's take the longer, roundabout way, and start with closer scrutiny of the mind. We'll get to the body later.

Frank: Yeah, that sounds right to me. We can start from the idea that the mind is not an immaterial substance in causal interaction with the body. I accept the arguments that show that we neuroscientists are wrong to suppose that the mind just *is* the brain. I'll go along with that. To have a mind, Aristotle showed us, is to have and to exercise the intellect and the will. But where do we go from there?

Richard: We need to examine the idioms that occur in association with the mind. For that will shed light on the logical character of our discourse concerning the mind and the exercise of our rational faculties. We speak of *having a thought in mind*, and of *having a thought at the back of our mind*. We speak of *making up our mind*, and of *changing our mind*, of *having half a mind to do something* or of *being in two minds whether to do something*. We call *things to mind*, and things sometimes *come to mind*. We *bear things in mind*, *hold things before our mind*, and sometimes things *slip out of our mind* or *lurk at the back of our mind*. We may *turn our mind to a task*, we may have a *sharp mind* or a *dirty mind*, an *active mind* or an *idle one*, a mind that …

Frank: [*waving his hand and groaning*] Enough already, Richard! You're a walking Thesaurus. Why is all this idiomatic stuff interesting? It's just English idiom. You won't find all this in other languages. So what's the big deal?

Richard: Well, I'm not making a big deal out of it, old chap. But it is interesting. And the fact that these idioms don't occur in French or German, or don't occur in the same way, doesn't mean that in a language in which they do occur, their occurrence may not teach us something important. Certain turns of phrase often express emblematic pictures embedded in our languages. The fact that they are commonly idiomatic does not mean that we cannot learn much from examining them. As Wittgenstein remarked, there is a whole mythology laid down in our language.[1] In many cases, our forms of representation exhibit the mythology in our language.

Jill: I don't know what you mean by 'form of representation', Richard.

Richard: A form of representation is the constant form in which we represent certain kinds of things, a form which may well be at odds with the logical character of what is represented. When we speak of *having a pain* in our foot or of there *being a pain* in our foot, as opposed to speaking of our foot's *hurting*, we represent pain in the form of an object. But, of course, pain is not an object of any kind. Similarly, when we present the number of things in nominal form, as when we say that the number of moons of the earth is one, we approximate representing numbers in the form of objects. Then, like Frege, we're misled into characterizing numbers as abstract objects. But they're not objects of any kind. We represent the mind as a possession, as something we *have* or *possess*, for we speak of *having a mind*, and of *losing* one's mind. We speak of knowledge as something we *acquire*, which we may *pass on to another* and *share* with another or *keep* to ourselves. We speak of memory as *a storehouse* in which we *keep* information and where we *store* our memories. We present thinking in the form of an *act* or *activity*, as when we order someone to think, or tell another not to interrupt one's thinking.

Frank: And why is that interesting?

Richard: Well, this mythology is often misleading. It is, so to speak, a form of linguistic iconography, and needs to be interpreted and understood – just as Renaissance iconography needs to be understood before one can see many Renaissance paintings aright. For example, you probably know that the figure of a man carrying a fish is almost always a representation of Tobias. And you might ask yourself why he is usually holding the hand of an angel.

Frank: No, I didn't know that. But what's the answer?

Richard: [*chuckles*] There isn't one. You've asked the wrong question.

Frank: Why? What's wrong with the question?

Richard: The question to ask is not why Tobias is holding the hand of an angel. The correct question to ask is why the angel is holding the hand of Tobias.

Frank: You've lost me. I don't follow.

Richard: It's the wrong question because there is no hidden story to the painting. Tobias is only there to identify the angel as the arch-angel Raphael, just as the fish is there to identify Tobias. Raphael is mentioned only in the book of *Tobit* in association with Tobias. So the only way to identify an angel as Raphael is by portraying the angel in the company of Tobias. And that may be crucial in identifying the guild or fraternity that commissioned the painting, for the archangel Raphael may be their patron.[2]

Jill: Brilliant. And correctly identifying and interpreting the linguistic iconography embedded in our language – the mythology laid down in our language – can similarly prevent us from asking the wrong questions and jumping to the wrong conclusions.

Frank: That's great. Linguistic iconography! Haha! Fantastic! Now how do you bring all this to bear on the misleading idioms of the mind?

Richard: Well, I suppose it calls for very careful and imaginative ana-lysis of both idioms and of other forms of our talk about the mind.

[*A small man, with twinkling eyes and brown/ginger hair, who has come up behind the three disputants, interrupts*]

Alan White: Of course you're right, Richard.

[*Richard jumps to his feet*]

Richard: Hello Alan. I'm so glad to see you. You know Jill and Frank, don't you? [*They exchange greetings*] Would you like to join us? You're the very person we need to help us with our enquiries.

[*Alan smiles and sits down*]

Alan White: Well, I'd be delighted to join you, and if I can contribute to your discussion, I should be very glad. [*Richard pours him a glass of wine and gives it to him*] Thank you. [*He takes a drink*] That is excellent … Now, I couldn't help overhearing what you were saying, Richard, and it seemed to me to be absolutely right. We speak of acquiring know-ledge, as if it were a possession, and we say that we can pass our know-ledge on to another or keep it to ourselves. That is the form in which we represent it. But just look at the way the word 'knowledge' and its cognates are used. Coming to know something is like becoming able to do various things. If I know something to be so, then I can answer questions on the matter, I can act on the information I previously learnt, and I can volunteer the information when it's requested or needed. So knowing is more like being able to do something than it is like possessing a piece of property. Bear in mind that when we share our knowledge, as opposed to sharing our possessions, we don't have any less; when we pass knowledge on to another, we also keep it. Similarly, we think of memory as a store of knowledge, but we need to recollect that when we remember something, we do not store it anywhere, rather we retain the information we acquired. Retention is not the same as storage. We *retain* abilities, we do not *store* them – there is no such thing as *storing* an ability. You can store your bicycle in the garden shed, but what on earth would it be to store your ability to ride it? Now, although knowledge is not a generic ability, it has a kinship with abilities.

Frank: So all these idioms are systematically misleading! We need to get away from these ordinary ways of talking and introduce a more scientifically accurate way of talking.

Jill: No, not at all – at any rate not if what we are trying to do is to clarify our existing conceptual scheme. You can, to be sure, introduce new terminology and new concepts. But that will not shed light on the conceptual problems that arise out of our existing concepts – including the conceptual problems about the mind-body relation that concern us now. If you want to understand how birds fly, you don't want to be given a blue print for building an aeroplane. If you are puzzled about our thermal idiom of heat and cold, warm and cool, as Berkeley was, you will not have your puzzlement cleared up by introducing the concept of temperature.

Frank: But these ordinary idioms are very misleading! You just showed how misleading they are.

Alan White: Well, they're only misleading once we start to think about them. In daily use, there is nothing misleading or bewildering about them at all. You aren't puzzled by the phrases 'I've made up my mind', 'I can't make up my mind' or 'I've changed my mind'. And you can readily explain what they mean. 'I've made up my mind' means 'I've decided', 'I can't make up my mind' means 'I can't decide' and 'I've changed my mind' means 'I've revoked my decision'.

Frank: Well, that's clear enough, but how does it help us? I mean, that's all obvious.

Alan White: Of course it's obvious. We're just reminding ourselves of some of the common or garden uses of the word 'mind'. What's not obvious at all is how to arrange these examples in a manner that will shed light on the problems.

Richard: Yes, we must find a way of ordering them so that they will give us a clear overview of the conceptual terrain – what Wittgenstein called 'a surveyable representation'.

Alan White: Exactly. Now if you reflect on the idioms associated with the word 'mind', you will find that a large number of them are clustered around four key concepts: first, intention and the will – which is what we were just looking at: making up one's mind, changing one's mind and so forth; secondly, memory and recollection; thirdly, thought and thinking; and finally, belief and opinion.

Frank: Could you spell that out? I'm not sure what you're driving at.

Alan White: Of course. What I mean is that many of these idioms can be arranged into four groups. Let's continue with intention and the will. In addition to the examples we've mentioned, we speak of *having it in mind to do something,* or of *being minded to do something* – which amounts to being inclined to do it or to intend to do it. We speak of *having half a mind to do something,* that is: we're tempted to do it. And we speak of *being in two minds whether to do something,* which means the same as: we're undecided. All right so far? [*The others nod*] Good.

Now, another range of idioms concerns memory and recollection: *to hold something in mind* is to retain it in memory; *to keep something in mind* is to ensure that one will not overlook it; *to bear something in mind* is to be aware of information acquired; *to call* or *bring something to mind* is to recollect it; *to cast one's mind back* to an incident is to try to recall something and, of course, *to be absent-minded* is to be forgetful or inattentive; for something to *slip, go* or *pass out of mind* is for it to be forgotten or overlooked.

Frank: That's neat. I guess that's the sort of thing Aristotle meant.

Jill: Yes. Yes, it must be. Do go on, Alan.

Alan White: All right. Let's turn now to thought and thinking. We speak of a *thought's crossing our mind,* which is just to say that something or other occurred to us. *To have something on one's mind* is to be preoccupied with it, and, to be sure, *to have a load taken off one's mind* is to be relieved from thinking about it. *To turn one's mind to something* is to start thinking about it, and to *focus one's mind on something* is to concentrate on it, to think hard about it. By contrast, when *one's mind wanders* one has lost concentration, and when *it goes blank* one does not know what to think or say. Of course, the idioms of cogitation also involve the venerable term 'idea'. For *to have an idea come to mind* is to have thought of something, *to have an idea lurking at the back of one's mind* is to have an incipient thought and *to have an idea flash through one's mind* is for something suddenly to have occurred to one. *To have an original cast of mind* is to display originality in one's thought, just as *to have a powerful, agile, subtle or devious mind* is to be quick, ingenious and skilful at problem solving, and for one's solutions, plans and projects to display subtlety and cunning.

Frank: [*laughing*] Enough, enough. I think we've gotten the point.

Alan White: One more observation, and then I'll stop. Note the number of idioms concerning the mind that are focussed quite generally on the faculty of rationality. For that gives us an important clue. Someone *is of sound mind* if he retains his rational faculties; someone is *out of his mind* if he thinks or proposes or does things that are irrational; someone is *not in his right mind* if he is distraught and one is said to *have lost one's mind* if one is bereft of reason and rational faculties.

Frank: OK. That sure is comprehensive, but where does it get us?

Richard: Well, for one thing, it surely confirms that however much we bungle our philosophical reflections on the mind, English idiom is on the side of Aristotle.

Frank: Would you spell that out a bit?

Richard: Isn't it obvious, old boy? All the idioms that Alan has placed on the table and ordered so neatly are concerned with rational faculties of intellect and will possessed by human beings, not with activities or properties of immaterial substances or of brains. It is human beings, not minds that think and reason, hope and fear, not minds. And it is human beings that deliberate, weigh reasons for thinking or acting and decide, not brains. All the idioms Alan has mentioned are linked with rationality – with the rational *psuchē* as Aristotle said, rather than with consciousness, or experience, or so-called conscious experience, as Descartes thought.

Alan White: I agree. But there are further morals to be learnt from our ordering of idioms. First, it is clear that each such idiom is easily paraphrased by a form of words in which there is no mention of the mind.

Jill: Doesn't that show that there is no such thing as a mind, that we don't really have a mind?

Frank: Yeah, what it shows is that the mind is just a fiction. It's not identical with an immaterial substance, and it's not identical with a material substance like the brain either, because it's not identical with anything!

Alan White: No, no. That's much too fast. Talk of the mind is not like talk of Father Christmas. Father Christmas is a fiction, but there is nothing fictitious about the mind. When we say that someone made up his mind, we're not saying something false or truth-valueless. It's not like saying to the children that Father Christmas will come in the night. When we speak of the mind, when we say that some great deed is a triumph of mind over matter, we aren't talking of some fictitious stuff, like the Elixir of Life, and we're not talking of some fictitious object, like the Philosophers' Stone, either. If the mind were a fiction, then *no one would have a mind* – [*he chuckles*] but to be mindless is to be stupid, and surely not everyone is stupid!

Richard: But one might, I suppose, say that all talk of the mind is a *façon de parler*.

Alan White: Well, I wouldn't say that, you know. For that too could be misleading. But one might say that our talk about the mind has a potentially misleading form in as much as we are prone to misconstrue it on the model of other quite different expressions that share the same form but have quite different uses. Our talk of the mind, unlike our talk of the brain, is not talk about a special kind of entity, but rather talk about an ordinary entity with special kinds of powers.

Jill: What entity?

Alan White: Human beings, of course. We're talking of human beings with powers of intellect and will, these being rational powers unique to language-using creatures. We are not, to be sure, *homo sapiens*, but we certainly are *homo loquens* – we never stop talking! [*He laughs*] Only creatures that have mastered a reasonably developed language can be said to reason, to apprehend things as affording reasons for thinking, feeling or doing, to weigh reasons and to come to conclusions on the basis of deliberation, and to do things for reasons.

Jill: Do you mean that other animals don't have minds? Isn't that going a bit far, Alan? I mean: they have conscious experiences, and there are well-known scientific experiments that show that dolphins and elephants are self-conscious beings too. It has been shown that they can recognize themselves in a mirror. So they obviously have a sense of themselves and can think of themselves. Surely that implies that they have a mind?

Alan White: I think that the matter needs careful scrutiny, Jill. It's far from obvious that being able to recognize one's face in a mirror betokens self-consciousness or a sense of the self – whatever that might be. Does the ability to recognize one's hand in a mirror also betoken self-consciousness?

Jill: No, of course not. That seems silly.

Alan White: Well, then, what difference does the face make? Why should the ability to recognize one's reflection in a mirror imply that one knows who or even what one is, let alone imply self-consciousness?

Richard: Let's not digress. I think we should finish our business with linguistic analysis.

Alan White: All right. Let's go back to where we were. Richard suggested that our talk of the mind is merely a *façon de parler.* One way in which that suggestion may be misleading is that it intimates that when we are speaking of the mind in our rich array of idioms, we are speaking of nothing. And that would be quite wrong. We are not speaking of *nothing*, rather we are not speaking of a *thing*.

Richard: So we might borrow Wittgenstein's remark about pain and say that the mind is not a something, but it isn't a nothing either.

Alan White: Perhaps. And you can bring out the fact that we're not talking of *a something* if you reflect on the fact that when we say that a man changed his mind, has a dirty mind and has turned his mind to something new, we are not say that there is something that is at once changed, dirty and has been turned![3] Of course, we are, in another sense, talking of the same thing, namely of the same human being. But we are, from case to case, saying something very different about him and the character and exercise of his powers of thought and will.

Frank: OK. I can see that. But what I'd like to know is this: does this linguistic analysis solve the mind/brain identity issue? Until we had that great discussion with Aristotle, I was convinced, like most of my neuroscience colleagues, that the mind just is the brain. But Aristotle showed us that that's bullshit. Does this examination of our mind-language get us any further?

Jill: Well, obviously. To be sure, if the mind is not a kind of thing, then it cannot be identical to a kind of thing. So it can't be identical to the brain. You can't have an array of powers between your ears, even though you wouldn't have those powers but for the stuff between your ears.

Alan White: That's right. But don't short-change the reductionists, Jill. The argument does not show that mental states are *not* identical with brain states.

Frank: Well now. If they *are* identical, that *would* satisfy most of my colleagues! And I'd feel OK with it too.

Richard: [*laughs*] Don't chortle too soon, old boy. There are plenty of other arguments to show that there is no identity here.

Frank: Ah! I was afraid you'd say that. So tell us.

Richard: Well, very briefly then – I don't want to digress from our main concern. First of all, for two state descriptions to signify one and the same state, the state descriptions must be describing states of the same thing. But neural states are states of the brain, not of the human being whose brain it is: if someone's brain is sclerotic or oedematous, it does not follow that the human being is sclerotic or oedematous, any more than if your heart is beating fast, then you are beating fast.

Secondly, the very notion of a neural state is ill-defined and underdetermined. We have no criteria of identity for a brain state that could determine what dynamic configuration of neurons, what neural activity, might be identical with feeling afraid of global warming or feeling overjoyed that Manchester United won the Cup.

Thirdly, no brain state could have either the grounds or the consequences of a mental state. There are reasons for feeling afraid of something frightening, but no reasons for being in a given brain state, whatever that may be. Similarly, if one believes that things are thus-and-so, that commits one. After all, you can't intelligibly say 'I believe that things are thus-and-so, but whether they are is still an open question for me'. But one can say 'I am in brain state B, but whether things are thus-and-so is still an open question for me'. So believing can't be a brain state.

Frank: Wow! That's a bit quick. I didn't follow all of that.

Richard: Well, you'll just have to try to think it through later, old boy. We have to pursue our quarries: the mind and the body.

Alan White: Good. But let me make one further point before we turn to the problems about having a body. It is evident from our linguistic descriptions that the mind is not the brain. But it is also evident that it is not the mind that thinks and feels, or imagines and reasons, it is the human being. I mean, no one would say 'My mind is thinking it over', 'My mind reasoned that things are so', 'My mind is getting angry' or 'My mind imagined that there was someone downstairs'.

Jill: Do you mean that these are not things our mind does, but things we do *with* our mind?

Alan White: No. On the contrary. We don't think or reason *with* anything – except perhaps with a pencil in hand.

Jill: But surely we *use* our mind when we think.

Richard: Yes, but not as we use our legs to walk. The mind is not an organ of any kind.

Jill: But we do say to each other: 'Use your mind'.

Richard: Yes, but that is not like 'Use your left hand'. It just means 'Think!'

Alan White: Exactly. The mind is not an aethereal organ, any more than it is a material organ. Remember that we also say 'Use your brains!', and that too means no more than 'Think!'. After all, the brain, unlike the hands or legs, is not an organ that we can move at will. Indeed, it is not an organ with which we can *do* anything, even though without it, we can do nothing. 'Use your head!' may be literal advice to the panting wrestler, but it is only metaphorical advice to the stupid schoolboy.[4]

Frank: So what *can* the mind do? You make it sound as if it is useless! If it is so feeble, we might just as well do without it.

Alan White: [*chuckles*] No, no, Frank. You've got hold of the wrong end of the stick. To have a mind is to have an array of rational faculties. Even in our curious possessive form of representation, we do not present the mind as an *agent*. But it is interesting that we sometimes speak of it figuratively as if it were a *patient*. After all, we say that our mind is in turmoil, – that is: that we are bewildered and confused; and that it is flagging – that is: we are weary of thinking or concentrating. We say that our mind has gone blank, that is: we do not know what to think or say; or that it is wandering, that is: we can't concentrate. But in this figurative form of speech, we don't present these as acts or activities of the mind, but rather as things that happen to it – as passivities, as it were. On the other hand, when it comes to thinking and reasoning, to deciding and forming intentions, to feeling overjoyed or excited – these are things done by a human being or responses of a human being, and we do not present them as activities of the mind.

Jill: To be sure, that sounds very convincing, and it all confirms Aristotle's conception. But I don't see what light it sheds on *having a body* and on what kind of relation *that **is***.

Alan White: Well, it's very suggestive.

Jill: In what way?

Alan White: It suggests a line of analysis we might pursue, and a method we might exploit. For we have now made it clear that *having a mind* is not standing in the relation of *having* to an object called 'the mind'. Moreover, *having a mind* is not possessing any *thing*. Now, it seems at the very least plausible that *having a body* is not standing in a relation of *having*, that is: of *possessing*, to one's body. So surely we should apply our linguistic analysis to phrases and idioms associated with the possessive form of representation characteristic of our discourse about our bodies.

Jill: But there's no analogy here at all. We have to grant that the mind is not a thing or entity of any kind, that it is not a substance, neither material nor immaterial. But it is obviously different when it comes to the body. For after all, the body *is* a physical thing, a spatio-temporal continuant consisting of matter. I *own* my body – it's *mine*, and I may be proud of it, or ashamed of it, feel comfortable with it or uncomfortable with it. It's obvious that I stand in a relationship to it.

[*Peter Strawson strolls in from the trees*]

Peter Strawson: Yes, of course, you do.

Richard: Hello, Peter! How delightful to see you. Do join us in our discussion. I think you know everyone here. [*Strawson nods and greets the others*]

Peter Strawson: Hello, hello. What a charming gathering. I should be very happy to join you.

Richard: Will you have some wine?

Peter Strawson: Thank you. That looks a very inviting bottle of *Arcadian*. [*Richard pours Strawson a glass. Strawson seats himself*] Good. Now, I overheard you talking about the notion of *having a body*, and of the relationship between a person and his or her body. An interesting question.

Jill: Yes, and it's one you've written about, if I remember correctly.

Peter Strawson: You're absolutely correct. I discussed the matter in *Individuals*.[5] I tried to clarify the relationship between a human being and his, or her, body by reference to the dependence of perceptual experience upon facts about the body.

Frank: What d'you mean, Peter?

Peter Strawson: Well, it is obvious enough if you think about it. Whether one sees anything at all depends on whether one's eyes are open. And what is visible to one depends on where one's body is located, on the direction in which one's head is turned and on how one's eyeballs are oriented. Now surely, such facts of experience explain perspicuously why a particular body should be spoken of as standing in some special relation – which we call 'being possessed by' – to the human person to whom states of consciousness are also ascribed.

Jill: Yes, that's exactly right. I possess my body. And I have inalienable rights over it.

Peter Strawson: Well, perhaps, perhaps. But at any rate such facts surely explain why a subject of experience should have a very special regard for just one body, why he – or she – should think of it as unique and perhaps more important than any other. Indeed, these facts might even be said to explain why, granted that I am going to speak of one body as *mine*, I should speak of *this* body as mine.

Richard: I'm not sure what you mean, Peter. Might you have spoken of some other body as yours?

Peter Strawson: Well, the experience-dependence of a human being on his location, orientation and the condition of his sense organs explains why a subject of experience should pick out one body from among others, give it, perhaps, an honoured name and ascribe to it whatever characteristics it has.

Richard: But Peter, that can't be right. A subject of experience does not *pick out* one body from among others to call his own. One surely has no choice, and could have no choice, in what to speak of as 'my body'. Nor could one be mistaken as to which body is one's own.

Peter Strawson: But can't we imagine that what visual experiences we have might depend on whether body A has open eyes, body B's head is turned in such and such a direction and body C is located in such and such a place?

Richard: No! No, we can't. We cannot tell a coherent tale within any such framework, Peter. For who is to duck when a snowball is thrown directly at one of these beings? Who is going to be hurt when the snowball hits him? And who will rub the point of impact and say 'Ow!'? I mean that the very notion of a subject of experience, and indeed the notion of agency, would disintegrate in the envisaged scenario. Our concepts of a sentient individual, of a living substance, of an agent, of a human being and of a person would crumble. For they would get no grip.

Peter Strawson: Ah … Good. Yes, I see my mistake. I have to withdraw that tale. Oh dear, what a blunder.

Frank: OK. I've been following your arguments, Richard, and I think I understand them. And it's not just that the concepts would be defunct, we'd have to abandon everything we know about biology and neuroscience.

Jill: But where does all this leave us? What bearing does it have on a person's ownership of her body?

Richard: Well, doesn't it make the whole idea smell rather bad?

Peter Strawson: As Wittgenstein used to say, it's easier to smell a rat than to catch one. [*Chuckles*] So what do *you* think is wrong, Richard?

Richard: Well, the body a human being *has* is surely not a body that he literally *possesses*. I think we can learn here from Aristotle. In the first place, he showed us that *having* a mind is not possessing some thing, but having an array of rational faculties, and that *having* faculties is not a matter of ownership. And bear in mind his remarks in the *Categories* about *having*.

Jill: You mean the incomplete chapter that upset him so much? When he rushed off to see Theophrastus and Neleus?

Richard: Yes. In the paragraph that is left – the one we know – he remarks on the peculiarities of the possessive form of representation. After all, as he noted there, having a wife or a husband is not owning anything, but being married to someone. Having a birthday is not possessing a birthday, and having a train to catch is not a form of ownership. Neither is having a body.

Jill: Why not?

Richard: Well, just reflect. One may have a house, a ticket to the opera or a copyright. These one can sell or give away. Once one has sold them or given them away, they are no longer one's own. They belong to someone else. Now, one *can* sell one's body. But doing so will not leave one bodiless or disembodied [*he chuckles*]!

Peter Strawson: Yes. Just as when Faust sells his soul to the devil, he is not left soulless.

Frank: Yeah, to sell one's body is to sell sexual services. A prostitute sells her body.

Richard: To sell one's body does not leave one bereft of a body or in possession of some other body.

Jill: But we speak of *having* two arms and two legs. Or of *having* a kidney, and these days one *can* give one's kidney away or indeed sell it. Isn't having a body just like that?

Richard: No, obviously not. I can lose an arm or a leg, but I can't lose my body.

Frank: [*chuckling*] But someone may find my body!

Alan White: [*laughing*] I agree, but then he needn't bother returning it. [*They all laugh*]

Peter Strawson: All right. All that is convincing enough. But why should we not say that the relationship between a human being and his body is nevertheless one of ownership, but that unlike the hum-drum relations of ownership you have been rehearsing, it is one of *inalienable* ownership?[6]

Alan White: Surely not, Peter. There is indeed such a thing as inalienable ownership. Something is inalienably owned if one legally *cannot* alienate it. In such cases, we know what it *would be* to alienate the thing. But if it *makes no sense* to alienate something – if we have no idea what would *count* as alienating it – then it cannot be inalienable either.

Jill: I don't see that.

Alan White: Look, Parliament may pass a law declaring that some property or some right is inalienable. But it can do so only if we know, have some conception of, what it *would* be to alienate it – if we know *what it is* that we are legally unable to do. But in the case of one's body, contrary to what the dualists contend, we have no idea what it would be not to have a body. I suggest that we can make no sense of a disembodied spirit.

Frank: Yeah. Nor of a disembodied mind.

Peter Strawson: Nevertheless, that which one calls one's body is, at least, a body, a material body. It can be picked out from others, identified by ordinary physical criteria and described in ordinary physical terms. And I do speak of it as mine.

Alan White: Peter, I am afraid to say that I think that you are confusing and conflating the body you are with the body you have. [*He chuckles*] As Thomas Moore said in Robert Bolt's play, 'I trust I make myself obscure'.

Peter Strawson: You do indeed, Alan. Very obscure.

Alan White: [*he laughs*] Look, the material body, the animate spatio-temporal continuant made of flesh and blood that *can* be picked out by others (but not by me), that can be identified and described in physical terms, *is me*. I am this living being [*he pats his chest*] before you. This living organism is not something I have, it is something I am. I don't call that which I am 'my body'. In one sense of the term 'body', namely a living organism, I am indeed *a* body. But I don't *have* or *possess* that which I am. I don't *have* myself and I don't *possess* what I am.

Frank: Now you really have lost me. What d'you mean: the body I am and the body I have?

Alan White: [*laughing*] The distinction was drawn in these terms by an acquaintance of mine in Oxford.[7] It's a good dictum, but, of course, it's deliberately both suggestive and provocative. You must admit, Peter, that the question: 'Which body is my body?' has little use. Science fiction, fairy tales, dreams of death and similar far-fetched cases apart, it's useless. Pointing at myself and saying 'This is my body' certainly has no use in picking out my body from among others, although it might be used to explain the phrase 'my body'.

Peter Strawson: But I can point reflexively [*he points at himself*] and say 'My body is aged', and point at someone else's body and say

'Her body is beautiful'. Don't I thereby pick out my body or her body from among others and ascribe to it whatever characteristics it has?

Alan White: Well, you point at yourself, and ascribe somatic characteristics to yourself. The body I am is *me* – this animate, spatio-temporal continuant made of flesh and blood. Surely no one is going to say that I own myself.

Jill: Well, no. What we want to say is that we own our bodies.

Richard: But surely, Jill, it's now obvious that our talk of having a mind has nothing to do with possessing something. It is just the form of representation we have for talking about our powers of intellect and will, and their exercise. All our talk of *having a good mind*, or *having a mind like a razor, having a devious mind* or *a subtle one*, is just a way of talking about human beings and their rational capacities. Now, may it not turn out that our talk of having a body is likewise a picturesque way of talking about ourselves – that is about human beings and their properties.

Jill: What do you mean 'and their properties'? We're talking about their bodies.

Richard: Yes, but to talk about the body a person *has* just *is* to talk about their bodily characteristics. I mean, if you have a skinny body, then you're skinny, if you have fat body, then you're fat and if you have a frail body, then you're frail.

Jill: You can have a beautiful body without being beautiful.

Richard: What do you mean?

Jill: If someone has a beautiful body and an ugly face, one would not say that she is beautiful.

Richard: Yes, of course. But that merely shows that there is yet another use of the word 'body' – a use that signifies the torso and limbs as opposed to the head or face.

Peter Strawson: Yes, that is surely right. Remember that they used to talk of having one's head struck off one's body.

Jill: All right. But even if we grant that talk of the features of the body you have can be cashed in terms of talk of your bodily characteristics, that doesn't show that you don't have a body, or that you don't have rights over it. If you're frail, then you have a frail body. The body you have is frail. And if you're fit, then the body you have is fit. The paraphrase goes both ways, you know.

Alan White: Before we go any further, we might try to put some order into the kinds of things we ascribe to the body we have, so to speak.

Frank: Y'mean the way we did with phrases about the mind?

Alan White: Yes, exactly. It's clear enough, if you stop to think. After all, we have a variety of corporeal determinables: a physique or figure, physical or aesthetic appearance, degrees of fitness and health, sensations and surface properties. We might call these 'somatic properties'. When we use the possessive form of *having a body*, are we not speaking of our somatic properties?

Peter Strawson: Yes, of course. That is an excellent suggestion, Alan. We say that we ourselves, or other people, have a beautiful or ugly, graceful or ungainly body – which means that they are beautiful or ugly, graceful or ungainly. We may describe ourselves or others as having a powerful, muscular or athletic body, which is to say that we are physically powerful, have well-developed muscles and an athletic build. We speak of someone having a healthy body, or a frail and failing one – that is that they are healthy, or physically frail and failing. To say that our body is aching all over is to say that we are aching all over. We say that someone's body is sunburnt or covered with sweat, is blue all over or is scratched and bruised all over, which means neither more nor less than that they are sunburnt, covered with perspiration, are scratched all over and are blue with cold.

Frank: Or with paint, like the ancient Britons, haha.

Peter Strawson: Quite so, quite so. One category of predicate that we ascribe to the body we have, to use Alan's idiosyncratic terminology, is to describe the superfices of the body we are. Is that not right, Alan?

Alan White: Yes, indeed.

Jill: I don't see what this has to do with possessing my body.

Richard: Look, if our talk of having a body typically boils down to talk of our somatic characteristics, then the question of ownership or of rights can't even arise. You don't *own* your bodily features. If you have a beautiful or graceful body, you don't *own* your beauty or gracefulness – they are features you have, not possessions you own. If you have a healthy and fit body, you're healthy and fit, but you don't *own* your health and fitness. You have no *right* to your health and fitness, any more than you have a *right* to your frailty and weakness.

Jill: But we have attitudes towards our body. We may be proud of our body if it is fit and beautiful, or ashamed of our body if it is ugly and ungainly, flabby and fat. We may feel comfortable with our body or ill at ease with it. We admire the bodies of the young and are shocked at the bodies of the mutilated. To be sure, these are attitudes *to the body*, to the *body we have*.

Peter Strawson: No, no! We must take Alan's suggestion seriously, Jill. These are attitudes *to ourselves* and *to other people*, attitudes the objects of which are somatic characteristics. To be proud of your fit and beautiful body just *is* to be proud of the fact that you are fit and beautiful – or at least that your figure is beautiful. It is to be proud of your physical *features*, not of a possession you own. And to be ashamed of your fat and unattractive body just is to be ashamed of being fat and ungainly. These are attitudes towards yourself and your somatic features.

Jill: But surely it is because I own my body that I have rights over my body. It *belongs* to me. A woman has a right to an abortion precisely because it is *her* body that bears a foetus, and she has a right to decide whether to have a child or not.

Richard: But Jill, women's right to an abortion does not have to rest on the misguided idea that they *own* their body. It is women that are pregnant, not their bodies. And it is women that bear a foetus. It is not as if they own something and the thing they own bears a foetus. One can perfectly well defend women's rights to demand an abortion in certain circumstances without any reference to the misguided idea that they own their body.

Frank: Yeah. Just like you can argue for prohibiting the logging of the Californian redwoods without claiming that trees have rights.

[*A brief silence*]

Peter Strawson: Where does this leave the mind-body relation?

Richard: Well, it doesn't look very good does it? I mean, how can a set of rational capacities stand in a relation to somatic features? *The body I am*, the living being I am, does not stand in any *relation* to its somatic characteristics or to its intellectual powers, any more than the chair you are sitting on stands in any relation to its height or weight. *The body I have* is no more than my corporeal characteristics. And corporeal characteristics cannot intelligibly be said to stand in a relation to intellectual powers.

Jill: Wait a minute. Something is wrong here. If I am five foot eight tall, and weigh nine stone, then my body is also five foot eight tall and weighs nine stone. But if it is wrong to say that I am my body, then it seems that two different things compete for the occupancy of the same space. Surely you aren't going to tell me, Alan, that I and my body occupy *different spaces*, or that they occupy the *same space* either! So something is wrong.

Alan White: No, no, Jill. The only thing that is wrong is that you have forgotten that the phrase 'the body I have' and the possessive phrases associated with a person's somatic features are no more than the surface grammar of our language. They are ways of speaking of our corporeal characteristics, not ways of referring to a physical object, let alone one which we inhabit. One cannot intelligibly ask whether the body I am is as tall as, or weighs more than, the body I have. Nor can one intelligibly wonder whether the body I have is as fat or as athletic as the body I am. It makes no sense to ask whether the two bodies occupy the same space at the same time, since there is only one body – one living spatio-temporal material continuant, not two.

Peter Strawson: Yes, that must be right. One might borrow an analogy from my old friend Quine. We can compare the phrase 'my body' with the phrase 'my sake'. If you do something for my sake, you do something for me. But you don't do something for two different entities, me and my sake. And you can't raise the question of how my sake is related to me.

Frank: That's neat!

[Alan and Richard nod in agreement]

Jill: But you are wrong to claim that all uses of 'my body', 'her body', 'their bodies' are reducible to talk of somatic features. Don't we write in our wills 'my body should be cremated'. That's not a way of talking about our corporeal characteristics or somatic features, is it?

Alan White: No, it isn't. I don't think that I suggested that *all* such phrases are reducible in this manner, only those that bear on the apparent ownership of one's body and on the apparent relationship between the mind and the body. When we speak of how our body should be disposed of after our death, we are speaking of our corpse. And of course, your corpse is neither the body you are, since you no longer exist, nor the body you have, since if you no longer exist, you no longer have corporeal characteristics. After all, Jill, you don't *have* your corpse, do you. The phrase 'my corpse' does not mean 'the corpse I have'. It means 'my remains', that is, what is left of me after I die – the dead organism. But the fact that 'my body', 'her body', 'their bodies' also have a use to refer to corpses does not support the claim that to have a body is a form of ownership. All it means is that the word 'body' has multiple uses, which are, to be sure, related, but are not synonymous.

Frank: Yeah, of course. When Newton speaks of bodies he just means any material spatio-temporal continuant. But when biologists and zoologists speak of bodies, what they mean are organisms.

Peter Strawson: And when architectural historians speak of the body of a church, they are not contrasting a mind with a body, but the nave with the aisles, just as when we speak of the body of a document, we are speaking of the main part of the document as opposed to the preface and appendices.

Alan White: [*laughing*] And when we speak of some*body* or any*body*, we're speaking of people. And when in Scotland an old lady is said to be a nice old body that's just the same as when we say in England that an old lady is a nice old soul. Haha.

Jill: All this nitpicking detail is fine, but we're losing sight of the mind-body relation.

Richard: But Jill, it's now obvious that there is no such thing as a relation between a person's mind and their body. If we take care to bear in mind the distinction between the body I am and the body I have, it is clear that the mind/body question simply evaporates, disappears. For the mind a person has and the body that a person has are not kinds of things that can intelligibly be said to stand in a *relation* to anything.

Frank: Yeah. Asking how my mind is related to my body would be like asking how green stands to the value of a dollar.

Richard: Say that again.

Frank: Well, dollar bills in the USA are known as greenbacks. A one dollar bill is green. But the color of a dollar bill can't stand in a *relation* to the value of one buck. [*He chuckles*]

Peter Strawson: Yes. Very good. My mind and my body are not relata standing in any kind of relation. *A fortiori* my mind does not *possess* or *own* my body. And my body does not *have* my mind.

Alan White: That's right. What would my body do with a mind, so to speak!

Jill: Yes, but *I* have mind and a body.

Alan White: Of course you do. But *having* is not a *relation* between you and your intellectual powers or between you and your corporeal features. You are a human being, and human beings are endowed with rational powers, on the one hand, and somatic features, on the other.

Richard: [*getting up and stretching*] Well, that was grand. Now what about dinner at the Ambrosian?

Peter Strawson: What an excellent suggestion. [*He empties his glass and rises to his feet*] Yes, that was a very illuminating discussion. Thank you so much.

[*Frank, Jill and Alan get up too and they all walk off into the trees talking to each other*]

Notes

1   Wittgenstein, *The Big Typescript* (Blackwell, Oxford, 2005), p. 317 (p. 433 in original TS). He ascribed the insight to Paul Ernst. The same observation was made by Nietzsche.
2   See E. H. Gombrich, 'Tobias and the Angel' in *Symbolic Images: Studies in the Art of the Renaissance* (Phaidon, London, 1972), pp. 26–30.
3   A point nicely made by Bede Rundle, *Mind in Action* (Clarendon Press, Oxford, 1999), p. 26.
4   A. R. White, *The Philosophy of Mind* (Random House, New York, 1967), p. 90.
5   P. F. Strawson, *Individuals* (Methuen, London, 1959), pp. 90–94.
6   Ibid., pp. 97–98.
7   P. M. S. Hacker, *Human Nature: The Categorial Framework* (Blackwell, Oxford, 2007), pp. 268–84.

SUPPLEMENTARY READING

P. M. S. Hacker, *Human Nature: The Categorial Framework* (Blackwell, Oxford, 2007), chap. 9, sections 3–4.
A. J. P. Kenny, 'The Geography of the Mind' and 'Body, Soul, and Intellect in Aquinas' in *Essays in the Aristotelian Tradition* (Clarendon Press, Oxford, 2001).
B. Rundle, *Mind in Action* (Clarendon Press, Oxford, 1997), chap. 2
P. F. Strawson, *Individuals* (Methuen, London, 1959), chap. 3.
P. F. Strawson, 'Self, Mind, and Body' in *Freedom and Resentment and Other Essays* (Methuen, London, 1974).

SECTION 2

TWO DIALOGUES ON CONSCIOUSNESS

INTRODUCTION

The subject of consciousness has been on the philosophical agenda since Descartes. Neither the ancients nor the medievals had a word for consciousness. The first occurrence of this expression in English is at the beginning of the seventeenth century, when it occurs with the same meaning as the Latin *conscientia*, namely 'to be privy, together with another, to some information'. Only gradually did it acquire its current uses. The idea of consciousness was introduced into philosophy by Descartes when developing his revolutionary conception of the mind, which we have inherited. The mediaeval Aristotelians conceived of the mind in terms of our powers of intellect and rational will, of our sensitivity to reasons and ability to reason. To study the mind is to study the exercise of those powers in human behaviour. Descartes redrew the boundaries of the mental and shaped the modern conception of the mind. He characterized the mind in terms of what he called 'thinking', and conceived of thoughts as 'whatever takes place within ourselves so that we are conscious of it, in so far as it is an object of consciousness.' Thinking therefore encompasses not only thinking as ordinarily conceived, but also perceptual experience, imagination, feelings and emotions, deciding, intending and willing – in short, subjective experience conceived as the object of consciousness. This conception flowered in the hands of the father of British empiricism, John Locke, who conceived of consciousness as a form of 'inner sense' or 'introspection'. Consciousness was thereby assimilated to self-consciousness, conceived (or misconceived) as knowledge of how things are, subjectively, with oneself. The nature of the mental was indeed to be explained in terms of consciousness, and the study of the mind was thought to be possible primarily by means of inner sense; that is: not by means of the investigation of the exercise of our powers of intellect and will as exhibited in overt behaviour, but rather by means of introspection and introspective reports.

After a period of relative quiescence, interest in the subject of consciousness revived in the last quarter of the twentieth century. This was in part due to the demise of psychological and logical behaviourism, and in part to puzzles raised by the development of computers (can they think?) and

Artificial Intelligence (could we, in principle, create robots that are exactly like us in appearance and behaviour, and if so, would they have conscious experience?). A prominent factor in redirecting philosophy to the subject of consciousness was the brief flirtation with what was called 'functionalism' in philosophy of mind. Cognitive psychology developed in the 1960s in reaction to behaviourism that had, in effect, eliminated the mental from psychology. This new school of psychology advertised itself as an attempt to reintroduce cognition into psychology. What it really introduced was non-conscious computation 'deeply buried in the mind-brain' and beyond the reach of consciousness. Functionalism in philosophy developed along parallel tracks: the mind was akin to a 'black box', which received inputs from our sense organs, and produced outputs of behaviour. The mind was the black box that mediated causally between inputs and outputs by means of mental states that were functional states of an organism standing in reciprocal causal relations. This seemed to combine the merits of behaviourism (the coordination of perceptual inputs and behavioural outputs) without sacrificing the mind. Mental states, according to functionalism, just were functional states of the organism, realized or actualized in states of the brain.

However, the functionalist programme (never actually executed) aimed to *define* what were deemed to be mental states (such as being in pain, perceiving things, believing things and so forth) solely in terms of the inputs and outputs they coordinate and their causal interactions with other internal states. It did not take long for critics to notice that this conception of the mind and of the mental conspicuously excluded any mention of the felt character of the experiences that sentient creatures enjoy and undergo, experiences of pain and pleasure, hunger and thirst, seeing and hearing, feeling cheerful or sad, longing, craving and wanting – in short everything that, on the Cartesian/Lockean conception of consciousness, we are immediately conscious of in inner sense. Against the background of functionalism, it seemed perfectly intelligible to suppose that there might be creatures ('zombies') just like us in all behavioural respects, subject to the same inputs and yielding the same outputs, and having the same causal connections between the internal, non-conscious functional states – but without enjoying or undergoing any conscious experience whatsoever.

So consciousness, 'conscious experience', was called in to save our humanity from soulless functionalism, and to banish the fear that all other people might, for all we know, be mere zombies. In a seminal article written in 1974, 'What Is It Like to Be a Bat?', Tom Nagel argued that an experience is conscious if there is something it is like for the subject of the experience to have it. For is there not something it is like to be in pain, to feel joy or sorrow, to see and to hear? A subject of experience is conscious if there is something it is like for it to be that subject. For while there is nothing it is like for a brick to be a brick, and nothing it is like for an ink-jet printer to be an ink-jet printer, there is surely something it is like for a bat to be a bat, a cat to be a cat, for us to be humans and indeed for me to be me. *This* is the essence of consciousness, and it is this that makes all forms of sentient life unique in the universe in having features that are not reducible to physical properties. What it is like for an organism to have a given experience was denominated 'the subjective character (or quality) of experience'. Each subjective experience was held to have its own qualitative character– its distinctive qualitative 'feel'. The problem of explaining these phenomenal qualities, it was held, *is* the problem of explaining consciousness. For what characterizes any conscious experience are the distinctive qualitative characteristics that accompany it. Nagel's article sowed the seed of subsequent 'consciousness studies'.

This conception of consciousness rapidly infected cognitive neuroscience. It presented neuroscientists with a new puzzle, namely what are the neural correlates of consciousness (NCCs)? Can neuroscience explain the strange phenomenon of consciousness – the 'what-it's-likeness-of-experience'? Caught in a veritable maelstrom of philosophical confusions, it is small wonder that cognitive neuroscientists found themselves out of their depth. Confronted by conceptual problems dressed up in the guise of empirical ones, it is not surprising that they floundered. For no conceptual problem can be solved by experiments. Consciousness seemed a mystery. Moreover, new questions came upon the carpet: what is consciousness for? If zombies are conceivable, then what is the evolutionary advantage of having conscious creatures at all? Moreover, how does consciousness emerge from mere matter? Does it not seem a mystery that matter should be conscious? Consciousness is surely caused by the brain. But how can a physical organ produce something like consciousness? How can the brain be conscious? There seems to be an unbridgeable gulf between matter and consciousness.

This is the background to the following pair of dialogues on consciousness that take place in an Oxford Senior Common Room. The protagonists are Adam Blackstone, an urbane Oxford linguistic philosopher, who is the host; Christopher Cook, an American cognitive neuroscientist; Bruce Palmer, an Australian philosopher who is a member of the self-styled 'cognitive studies community'; Sandy MacPherson, a Scottish biologist; and Jocelyn Thomas, the guest of the Viennese stranger, who is visiting Oxford. As Plato allowed himself an Eleatic stranger, I thought that I might be allowed to introduce a Viennese stranger, familiar with the works of Wittgenstein, but not tied to them. His is the Socratic role of keeping the discussion on track and the Wittgensteinian role of disentangling the knots in the reflections and assertions of the others.

Third Dialogue

THE MYSTERY OF CONSCIOUSNESS

Protagonists:

Adam Blackstone: a middle-aged Oxford don; the host of the gathering; educated Oxford accent; dressed in a well-cut sports jacket and elegant tie, casual slacks; a linguistic philosopher.

Christopher Cook: a distinguished American cognitive neuroscientist from California; middle-aged; pronounced American accent; dressed in jeans, sports jacket and open necked shirt. His views are a synthesis of a large number of different views among cognitive neuroscientists.

A Stranger: late-middle-aged philosopher from Vienna; dressed in a dark suit, with a bow tie; pronounced German accent, coupled with a speech mannerism of punctuating his remarks with a slight 'Hmm' or interrogative 'uh?' A remote intellectual descendent of Plato's Eleatic stranger in *The Laws*. He is obviously familiar with the works of Wittgenstein.

Bruce Palmer: an eminent Australian philosopher, a member of the self-styled 'consciousness studies community'; in his mid-thirties; relaxed; amiable; pronounced Australian accent; dressed in T-shirt and jeans. His views are a synthesis of many different views among philosophers who have nailed their colours to the masts of cognitive science.

Sandy MacPherson: a Scottish biologist in his fifties; soft Scottish accent; dressed in corduroys, tweed jacket and woollen tie.

The scene is a small Oxford Common Room. The room is panelled. Oil paintings of past fellows and of presidents hang from the picture rails. There is a bust of Socrates on a stand in the corner. Lighting comes from the central chandelier and sconces on the walls. A fire is flickering in the fireplace. There is a well-stocked bookcase against one wall. There are five comfortable leather armchairs arranged around a low central table with books and papers scattered on it. On one side is a sideboard on which there are glasses and bottles of postprandial drinks. Bruce

Palmer, Christopher Cook, Adam Blackstone and his Viennese guest are helping themselves to drinks.

Adam: Well, gentlemen, if you all have your drinks, let's sit down and start our discussion. [*The others take their seats*] I asked you to dinner tonight so that we might spend an hour or so after the meal discussing that most vexatious subject – consciousness. And I've asked my old friend from Vienna to join us. He is devoted to philosophy, and will, I am sure, be able to shed some light on this subject. [*He sits down*]

Christopher: Well, y'know, you philosophers have had centuries to think about the subject, and you haven't come up with much. There are plenty of us working on the NCC – the neural correlates of consciousness – but philosophers haven't been of much use. Are you going to give us a philosophical lecture on consciousness?

Stranger: I should feel some shyness, Christopher, at the idea that, at my first meeting with your distinguished friends, instead of exchanging ideas in the give and take of ordinary conversation, I should spin out a long discourse to you by myself, as if I were giving a display of eloquence and authority. For indeed, the question you have just raised is not so easy a matter as one might suppose. The subject calls for a very long lecture, hmm? – perhaps indeed for two or three – but that would not be appropriate. On the other hand, to refuse a request, especially one put to me in such terms as you have used, hmm, strikes me as a breach of civility. I think we should have a conversation together, and I shall perhaps ask you some questions, uh?[1]

Bruce: I thought you were going to give us your theory of consciousness, and we were going to ask you questions.

Stranger: *Ach*, no. I have no theory of consciousness. And if I had, what good would it be to you? – We do not need a theory at all. What we need is clarity – and an overview of the conceptual domain. No, no. I shall ask you questions – and, as you will see, good questions in philosophy are worth far more than yet another theory. For the good question lets in the light, uh. Most philosophers go wrong before they have even begun, exactly because they have not asked the good questions.

Adam: That's fine, my friend. We'll get a discussion going, and you come in whenever you please. Chris, would you like to make a few remarks to start us off?

Christopher: OK. [*He pauses to collect his thoughts*] Look, it's generally acknowledged by cognitive neuroscientists that perhaps the greatest problem in all of biology resides in the analysis of consciousness.[2]

Francis Crick, the great molecular biologist, and his young colleague Koch, an eminent neuroscientist, held that consciousness is the most mysterious aspect of the mind-brain problem.[3] Frisby, a distinguished psychologist, said that consciousness remains a great mystery, despite considerable advances in our knowledge of perceptual mechanisms.[4]

Bruce: Sure. And it's not just scientists, y'know. Philosophers go along with it too. Daniel Dennett said that consciousness is the most mysterious feature of our minds.[5] My colleague David Chalmers wrote that conscious experience is the most familiar thing in the world and it's the most mysterious.[6]

Adam: Slow down, Bruce, slow down. That there are living creatures on our planet may be awesome, but surely not mysterious. That some of them are conscious, sentient, creatures is a great wonder, but not a great mystery. Don't confuse the awesome and wonderful with the mysterious. Volcanic eruptions are awesome; so are colossal waterfalls like the Niagara Falls, but there is nothing mysterious about them. Now, something that induces *conceptual bafflement* may *look* mysterious, but it isn't. 'There are no mysteries' might be a good motto to put on the coat of arms of philosophy.

Stranger: *Ja*, that's right. We project our puzzlement onto the phenomenon that puzzles, and then we create a mystery out of our own mystification. Hmm. *So*, now, what is this 'mystery'?

Bruce: Y'know, Thomas Huxley, Darwin's sidekick, put his finger on the matter in the nineteenth century. He pointed out that it is just baffling, amazing, that a state of consciousness should come about as a result of irritating nervous tissue.[7] That is just as unaccountable as the appearance of Djin when Aladdin rubbed his lamp! John Tyndall, a great nineteenth-century physicist and popularizer of science, said that the passage from the physics of the brain to the corresponding facts of consciousness is unthinkable.[8] And Ian Glyn, a distinguished Cambridge scientist, recently remarked that we haven't much more idea why consciousness should emerge from events in the brain than Thomas Henry Huxley did.[9] And …

Stranger: *Ja* …, I have got the point, Bruce. But although all these great men express their bafflement, it is not clear to me *why* they are so baffled, uh.

Christopher: I should've thought that's obvious. I mean, my present state of consciousness is a feature of my *brain*.[10] So the brain clearly has some property that confers consciousness upon it. We are all agreed that the brain is the organ of consciousness, OK.[11] Consciousness is an emergent property of non-specialized neurons,[12] a property of a neural network.[13] Now, it's just obvious that the gulf between brain states, as described by physiologists and neuroscientists, and mind states as described by human beings, is unbridgeable.[14] Antonio Damasio – I'm sure you've heard of him – Antonio Damasio said that there's a mystery *how* images emerge from neural patterns. He said, quite rightly, that how a neural pattern *becomes* an image is a problem that neurobiology has not yet resolved.[15]

Adam: What have images got to do with it, old boy? After all, when you see me now, you don't see an image, you see me. You see an image when you look at the paintings on the wall. But painted images don't emerge from the brain [*he chuckles*] – they emerge from a paintbrush. You can see a graven image of our patron saint over there [*he points at the bust of Socrates*], but that emerged from a block of stone.

Stranger: Christopher, let us go back a few paces, uh. Why do you want to say that the brain is an organ of consciousness?

Christopher: Well, because we wouldn't be conscious but for the normal activity of our brain.

Stranger: Hmm. Let us take this slowly, uh. We all agree that the organs of locomotion are the legs – you walk with your legs, don't you?

Christopher: Yeah, sure.

Stranger: And although you would not be able to walk with your legs if your brain were so damaged as to cause paralysis, nonetheless, you don't walk with your brain, uh?

Christopher: Yeah … Yeah, I guess so.

Stranger: Right. *So*, we walk with our legs, which are the organs of locomotion, and we see with our eyes, which are the organs of sight. We need a normally functioning brain to walk and to see, but the brain is not the organ of walking or of seeing, is it?

Christopher: Yeah, I see. No, it obviously isn't.

Stranger: So what makes an organ – like the eyes, ears, nose, or hands or legs, uh – an organ *of* something or other, an organ of sight or smell, of locomotion or manipulation?

Christopher: I guess it's an organ we *use* to do something.

Stranger: Exactly *so*, *Ja*! We put our eye to the keyhole to look through it, and bring our eyes closer to something to see it better. We sniff with our nose, and bring it close to the object we are smelling. We use our hands to manipulate things, uh, and we walk with our legs.

Christopher: OK.

Stranger: Well, then, is it not obvious that we do not use our brain to become conscious of things? Hmm. We cannot *do* anything with our brain. We walk by moving our legs, but we cannot act by moving our brain, because we cannot move our brain, uh? We look at and watch things with our eyes, but the brain is not an organ we can use to do *anything*. We are not conscious *with* anything. The brain is not *an organ* of consciousness, although we would not be conscious without a brain – we would be dead without a brain!

Christopher: But surely we think with our brain, just as we digest food with our stomach!

Adam: There's no 'surely' about it, Chris. Our stomach digests food, and so we can say, derivatively, that we are digesting our food. But it is not as if our brain thinks, and we can say *derivatively* that we think. We think with our brain only in the sense in which we love with our heart.

Christopher: What d'you mean?

Adam: Well, we say 'I love you with all my heart', and that just means, 'I really do love you', and similarly, we tell someone 'Use your brains!', and what that means is 'Think!'. Both are just figurative turns of phrase.

Bruce: OK. But none of this shows that it is not the brain that is conscious. Even if we grant you that the brain is not the organ of consciousness, it doesn't follow that it is not the brain that is conscious. I mean, all the neuroscientists and everyone in the consciousness studies community agree that it is the brain that is conscious.

Stranger: [*chuckles*] No, no. Never mind about the consciousness studies community, Bruce. A few decades ago the sense-data studies community agreed about all manner of strange things. I recollect a friend contemplating applying to NATO for a grant to go to the south of France to look for sense-data, because we were falling behind in the sense-data race![16] Haha! No, no, we must think this through for ourselves. *So*, tell me, Bruce, when do you say of a subject of experience that he or she is waking up, or regaining consciousness?

Bruce: When they open their eyes and become responsive to what they see and hear.

Stranger: And when do you say that a subject of experience suddenly became conscious *of* something?

Bruce: Well, when one observes the chap having his attention caught by something in his perceptual field – I guess. When he stops in his tracks and cocks his ear, so to speak, and listens carefully, and so on.

Stranger: But does the brain stop and listen carefully? uh? There is no such thing as a brain stopping in its tracks, since there is no such thing as a brain walking; the brain cannot cock its ear and listen carefully, since the brain has no ears; and the brain cannot turn its gaze and look intently at what caught its attention, since the brain has no eyes, uh? *Nein, nein* – it is not *the brain* that becomes conscious of something perceived; it is *the animal* whose brain it is that may become conscious of something. It is not the brain that is consciousness; it is the living animal. It is human beings who regain consciousness, who sit up and become conscious of things. Hmm! Of course, they would not do this if not for the appropriate brain activity. Consciousness *is a property of sentient creatures,* no?

Bruce: OK, but it's still a mystery how consciousness can emerge from neural activity, isn't it? It still seems to me that Huxley was right to say that it's remarkable that a state of consciousness, a visual experience, can come about as a result of irritating nervous tissue. And Tyndall was right to say that the passage from the physiology of the brain to corresponding facts of consciousness is just unintelligible – we just don't understand it, and don't know how to go about constructing a theory that would explain it. There just is a *gulf* – an unbridgeable gulf – between brain states and mental states of human beings.

Stranger: *Ach, nein.* – Slow down, Bruce, you are going too quickly? And also you are getting too, how should I say, er … *melodramatic.* There are no *gulfs* here.

Adam (*aside*) **:** Only gaffs!

Stranger: Tell me, Bruce, when someone drinks a bottle of whisky and passes out – loses consciousness – does that seem to you to be *a miracle?* Uh? If our friends become merry on drinking the fortified punch at the party, is that *mysterious?* Is this something that 'crosses the gulf' between the physical and the mental?

Bruce: No … I guess not. If someone were to drink a bottle of whisky and not pass out – that would be a miracle [*he laughs*].

Stranger: *Ja, ja* – that is correct. And we can explain perfectly well why alcohol causes us to become, as you say, first jocose, then morose, then bellicose and finally comatose [*he chuckles*]; isn't it so, Christopher, uh?

Christopher: Yeah, sure. We know all about the impact of alcohol on the prefrontal cortices.

Stranger: *Ja,* of course. My point is that we don't feel that anything mysterious has happened. *So* – again – if you have a brain operation, and the surgeon stimulates some part of the cortex with an electrode and you *seem to see* flashing lights, would you think that you have participated in a *mystery?* – uh? – It may be odd, but not a mystery.

Christopher: OK, OK. But where does that leave us? Isn't it mysterious that when light enters our eyes, and stimulates the light sensitive cells of the retinas, and impulses are sent along the optic nerves, across the chiasma to the visual striate cortex, and then somewhere along the line, you have *a conscious visual experience?*

Adam: You mean that when you are sitting in normal light in front of a bowl of red apples, and you look at the apples, then it is a mystery that you see red apples? Come now … I should have thought that what would really be mysterious is that you should look at a bowl of red apples and see green apples – or a white cockatoo!

Christopher: No, no, you're missing the point, Adam. It *is* a mystery, a mystery of the transition from something purely physical to something mental – *that's* the mystery.

Adam: I think you have got hold of the wrong end of the stick, Chris. It isn't in the slightest bit mysterious that when people get drunk they pass out. Nor is it mysterious that when you, a normal sighted human being, look at a red apple, what you see is a red apple. It only starts to *look* mysterious when we redescribe this mundane phenomenon in neural terms, and then switch from one level of description to another.

Christopher: You've lost me.

Adam: Look! We say, correctly, that when we look at a red apple in normal light, then light waves in the 630–700 nm wavelength are reflected off the surface of the apple onto our retinas. They stimulate the light-sensitive cones of the retina. These transmit neural impulses along the optic nerves via the thalamus to the primary visual cortex, where select groups of cells are responsive to colour, others to shape and yet others to motion. These in turn transmit further impulses to other parts of the brain.

Christopher: Yeah, sure. We all know that. But how does visual experience emerge from this – that's what we want to know. No one can deny that consciousness is caused by the brain, and yet we have no idea how this is possible.

Adam: But that is just because you are looking at it wrongly. Two different kinds of conceptual confusions generate the sense of mystery. The first is the thought that the brain has to bring all this information together again to form an image of a red apple. The second is the thought that somehow, miraculously, this neural activity becomes a visual experience – that there is a miraculous, mysterious, transformation of the physical – the physiological – into the mental.

Christopher: But it really is mysterious. – It transforms matter into mind.

Adam: [*laughs*] No, no! This is just a conceptual muddle, not a mystery, old chap. The source of the confusion is the supposition that somewhere at the end of the causal chain of cortical events that begins with the response of the photoreceptors to light falling on the retinas, we are going to find a visual experience. And we immediately go on to think that this visual experience, which is 'created by the brain', is mysteriously different from the sequence of neural events that give rise to it. *Ex hypothesi*, this visual experience occurs *in* the brain, so we think that it belongs *to* the brain – as if it were *the brain* that has visual experiences.

Christopher: [*hesitantly*] So … what's wrong with that? … I don't see what you're driving at.

Adam: The point is a simple one. However far you trace the neural connections in the brain that are involved in seeing something, you will *never* find a visual experience – only more neural connections. The visual experience will always seem *just beyond* whatever link in the chain you may have arrived at. Moreover, the fact that the cells in the primary visual cortex that are colour-sensitive are distinct from and differently located from those that are movement- or shape-sensitive does not imply that the brain has *to bring all the information together* in order to *form a picture*. After all, the brain doesn't *have* any *information* about the movement, shape or colour of things. To have, to possess, information is to know something. But brains cannot be said to *know* anything. And the brain does not *form an image* or picture of what we perceive. What we see is not a picture of an apple, but an apple. And what the brain does is make it *possible* for us to see the apple. It does not present us with a picture of an apple or of anything else. [*Silence*]

Look, imagine we could shrink you to the size of a red blood corpuscle, so that you could get into the arteries and go on an expedition through a living brain. You would walk through miles of blood vessels, look at endless neurons, observe countless dendrites, examine billions of synaptic connections. But *nowhere* would you observe an image, or see a visual experience.[17]

Christopher: Yeah … I'm beginning to see what you're getting at. So it's not that irritation of the cells of the retina produces visual experience. It's that irradiation of the retina by light waves enables *us* to see the things that reflected the light.

Stranger: *Ja*, precisely so. Adam has right – what you *see* is the apple in the fruit bowl, what you *have* is a visual experience. We talk, in philosophy, of *having* visual experiences, but that is misleading. For all our talk of visual experiences introduces *no new entities* that are not already present in talking of *seeing* things. The activity of the brain that is necessary for us to see things does not create any new, mysterious things called 'visual experiences', uh?

Christopher: Yeah, I see. The complex series of neural events that follow on the impact of light on our retinas are the neural processes that are *biologically necessary* for us humans to be able to see things in our immediate environment. So the visual system is the vehicle of our sense of sight, and its normal functioning is necessary for exercising our visual powers.

Stranger: *Ja*, that's correct – just as the cylinders and pistons are the vehicle of the horsepower of the car, and their normal functioning is needed for the use of that power.

Bruce: Well *I* don't see! I mean, *how does consciousness emerge from mere matter?* Wasn't it this question that baffled Huxley and Tyndall? Wasn't it this that struck them as an unfathomable mystery? I mean, it's consciousness that distinguishes us from mindless nature, isn't it? But how can consciousness emerge from mere matter? That's the big question!!

Stranger: But Bruce … *Can* consciousness emerge from *mere* matter, uh? [*He leans forward and slaps the table*] This is mere matter. Can consciousness emerge from this? Consciousness does not 'emerge' from mere matter at all.

Bruce: Ah, so you're just a closet dualist, after all! I thought so?

Stranger: [*much amused*] My dear Bruce, there are not sufficient 'or-s' in your story of either dualism or reductive materialism. – And I don't live in the closet. Haha. No, no, be serious. Hmm.

Bruce: OK. So consciousness doesn't emerge from mahogany. But it does emerge from neural tissue, from the brain. And that is just as mystifying as if it emerged from mahogany. I mean, the brain is a material thing – a physical system.

Stranger: *Ach nein!* Forget about *emerging*, Bruce! It is only confusing you. It makes you think of consciousness as an aura emerging from the skull. Hmm. Remember that living creatures are what we *contrast* with mere matter. They are not *mere* matter; they are living organisms, uh.

Bruce: OK, but the brain's a physical thing, a physical system. And how can a mere physical system have consciousness?

Stranger: But Bruce, mere physical systems cannot, as you put it, *have* consciousness.

Bruce: How's that?

Stranger: Well, just think what we can call 'mere physical systems'. The solar system is a mere physical system, and it does not even make sense to say of it that it is either conscious or unconscious, uh? The atmosphere is a mere physical system, and, of course, no one expects it to wake up or fall asleep. We can say of a computer that it is a mere physical system. But we don't think that our laptop wakes up when we turn it on, or that it falls asleep when we turn it off; uh? Mere physical systems cannot be said to *have consciousness*. Only *living*, biological, 'systems' – *living animals* – can intelligibly be said to have consciousness. And it is very misleading to refer to *them* as systems – for the word 'system' misleads here. Only sentient creatures can be said to lose or regain consciousness, or to be conscious of something in their field of perception. Is that mysterious? Hmm.

Adam: And living creatures don't *have consciousness*, they *are conscious* – or unconscious. Introducing a nominal here is misleading. Indeed, it is just a misleading Germanism, a bad literal translation of 'Bewusstsein haben' that emigrated to the USA and has now invaded our shores.

[*There is a knock at the door*]

Adam: Hello! Come in! [*Professor Sandy Macpherson comes in*] Ah, hello Sandy.

Sandy: Good evening. David told me that you were having a discussion about consciousness. Would you mind if I joined you? I have been bothered with a question about consciousness for ages, and I was hoping that you lot might be able to answer it.

Adam: Please come in, Sandy. Help yourself to a drink, and then come and sit down here [*he points to the vacant armchair*].

Sandy: Thank you [*he pours himself a whisky*] – can I top anyone up? [*He fills the glasses of Bruce and Chris*] And cognac for you two? [*He returns the bottle of Scotch, fills Adam's and his friend's glasses with cognac, returns the bottle to the sideboard and sits down in the vacant armchair*]

Stranger: *So.* May we continue? … *Ja* … hmm … Bruce, although animals are physical things, they are not mere physical systems. Indeed, they are not *systems*. Living beings are what we *contrast* with 'mere physical systems'.

Bruce: [*sarcastically*] You mean because they possess an *élan vital* or an *entelechy*.

Stranger: [*reproachfully*] *Ach*, Bruce! That is not thoughtful. – *Nein, nein.* Living beings are physical things, but not *mere* physical things. They can do things which no *mere* physical things can do. They metabolize from the environment, they grow, they reproduce. Animate beings are self-moving, they are sentient, they feel pleasure and pain, they pursue goals, uh. To call these 'physical systems', let alone 'mere physical systems', is not merely *eine*, how do you say, *barbarismus* – it misleads those scientists who, in the name of the scientific asepticism, do so.

[*Silence*]

Adam: What was bothering you, Sandy?

Sandy: Well, many biologists, and other scientists too, are puzzled and even embarrassed by the question of what consciousness is for.

Stranger: That is a very strange question? What do you mean: what is consciousness for?

Sandy: Well, y' know, this has become a widely debated question right across the sciences. Stuart Sutherland famously remarked that consciousness is a fascinating but elusive phenomenon, but that it is impossible to specify what it is, what it does, or why it evolved.[18] And Johnson-Laird, a well-known cognitive scientist, said that no one knows what consciousness is, or whether it serves any purpose.[19]

Adam: My dear Sandy, how could they possibly have any idea of what it is *for*, if they have no idea what it *is*? [*Chuckles*] As William James put it, that would be like looking in a pitch black room for a black cat that isn't there.

Sandy: Well ... But at any rate, the distinguished neuroscientist Sir Horace Barlow wondered what selective advantage is conferred on animals, including ourselves, by consciousness.[20] He put it rather well, I thought: Is consciousness just an epiphenomenon of brain processes – a sort of whining of neural gears? Or does it have a more important role in the survival of our species? And the great mathematician Roger Penrose also raised the matter.[21] The crucial question, he suggested, is whether consciousness actually *does* anything. Is it helpful to the creature that has consciousness, so that an otherwise equivalent creature, but without consciousness, would behave in some less effective way?

Stranger: *Ach so* ... *Ja* ... Very fascinating. And did any of these thinkers say what a creature that was like us only without consciousness might be like, uh?

Sandy: Well, I don't like investigating biological problems by means of science fiction, but their idea is that there could, logically speaking, be creatures they call 'zombies', who behave just as we do, only are not conscious. Now that seemed fishy to me ... Adam, isn't it fishy?

Adam: Nothing but pickled herrings, old boy! – I can see that someone who indulged in such absurd science fiction might well be baffled by the question of what consciousness is for. If there is a being that looks like us, that behaves in all respects like us, that responds and reacts to circumstances as we do, then to be sure, he *is* one of us! And he is conscious like us too.

Bruce: But that is just what the consciousness studies blokes deny. They say that zombies behave exactly as we do; only there is darkness inside, so to speak. They lack consciousness. [*He turns to the Stranger*] I mean, that seems perfectly OK to me. What's wrong with that hypothesis?

Stranger: Hmm. Well, let us take it slowly, huh? Suppose a zombie can find his way through an obstacle course without bumping into things, or falling into holes. Huh. Suppose the zombie can tell us what is before him in the light, but not in the dark. And suppose also that he can say what is before him when his eyes are open, but not when they are closed, uh. And suppose that he shows pleasure when he sees what is beautiful and is revolted when it is disgusting. *Ja.* Now, what sense can we make of the idea he cannot see a thing?

Bruce: Well, I guess the answer has to be that there is no inner light, so to speak.

Stranger: But Bruce, consciousness is not *an inner light*. If there is light anywhere, it is all 'out there', so to say, not 'in here' in one's head. Sentience is sensitivity to what is *outer*, so to say – not observation of anything *inner*.

Sandy: Well, that is really fascinating. But then why is the idea of creatures that behave just like us, only are wholly lacking in sentience, so persuasive? Why do we fall for it so easily?

Stranger: *Ja*, I agree that it is a very compelling picture. There are many reasons for this, uh. But the most powerful one comes from conflating consciousness with the ability to say what we are experiencing. For we are inclined to think that the ability to *say* how things are with us derives from the ability to *see*, by introspection or 'inner sense', how things are with us.

Bruce: I don't follow you.

Stranger: Look, my friend, we can say that we see something, uh? We can say that we are in pain, no? It is natural, *nicht wahr*, to think that our ability to do so comes from introspection – from looking inside to see whether we are seeing or not, and to see whether we are in pain, no? And we think of introspection as a form of inner vision, and of introspecting as a kind of looking *inside*, whereby we can see how things are with us mentally speaking. Hmm?

Bruce: Yeah, I guess that's right. I mean, how would we know if we couldn't introspect?

Stranger: But now think – what do we mean by 'introspect'? Hmm. What are you supposed to look *with*? With the eyes of the mind? Does the mind have eyes? And *where* are you supposed to look? In the brain? Uh. Do you look into *your brain* to see *what it is that you see*?

Bruce: No. No, that's obviously wrong. I guess that we introspect the mind.

Stranger: But is the mind a place? Huh? Is it a space? Huh? *Nein.* And when you say that you have a headache, do you have to look at anything?

Christopher: I see. Y' mean that one simply has a headache and says so.

Stranger: Correct. Hmm. The whole notion of looking inside is no more than, Hmm – *eine Bild* – a picture.

Bruce: What do you mean, 'a picture'?

Stranger: I mean that it is no more than an *emblematic representation* of our ability to *say* how things are with us.

Christopher: I see! So we confuse the emblem with what it is an emblem of. Yeah. That's good!

Stranger: *Ja.* We take the picture, the emblem that represents our ability to say how things are with us, uh, and we jump to the conclusion that we can say how things are with us only if we can look inside, so to say. And then we think that we can see what is going on inside only if there is some inner light, uh – otherwise we would not be able to see.

Adam: That's brilliant! – And then we indulge in science fiction and introduce zombies who are supposed to be behaviourally just like us, only without any inner light of consciousness. We fall victim to our own imagery.

Christopher: OK. That's very good. I find that persuasive. But now, if the science fiction of zombies is just nonsense, then what remains of the question of whether consciousness serves any purpose, or what its evolutionary role has been?

Sandy: Well, I don't think the answers were really dependent on the stories about zombies. Sir Horace Barlow suggested that the point of having consciousness is that it links the individual to the community in which he lives. Consciousness is impossible without social experience, he says. Suppose that consciousness is awakened in the infant by the first mutual communication with another person – perhaps the first smile returned … To enlarge this experience and bring it partly under control, the infant brain, Barlow said, must build a model of what it is interacting with. It must build a model of the mother and her brain which will tell the infant when smiles will be returned, and when other responses and interactions will occur. The crucial feature of consciousness, according to Barlow, is that it requires a remembered partner for its introspections: consciousness is taught, awakened and maintained by interaction with other modelled minds.

Adam: [*mutters*] Extraordinary!

Sandy: Well, you know, there is a Cambridge biologist, Nicholas Humphrey, with some very fishy views. He went down a very similar route. He said that the advantage to an animal of being conscious lies in the purely private use it makes of conscious experience *as a means of developing a conceptual framework which helps it to model another animal's behaviour*. Now this strikes me as bizarre, although I'm not quite sure why. Somewhere along the evolutionary path that led from fish to chimpanzees, Humphrey said, a change occurred in the nervous system which transformed an animal which simply behaved into an animal *that informed its mind of the reasons for its behaviour*. Now what on earth does that mean, Adam? Humphrey goes on to say that *the inner eye of consciousness* – that's his phrase, mind you – that the inner eye of consciousness provides us with an extraordinarily effective tool for understanding, by analogy, the minds of others like ourselves.[22] Well now, what do you think of that, Adam?

Adam: It seems to me to be profoundly confused. If consciousness only emerges phylogenetically with highly social animals, we must suppose that huge numbers of sentient non-social animals are not conscious or unconscious, do not fall asleep and later awaken, and do not have their attention caught by things they peripherally perceive and of which they become conscious. But it is absurd to suppose that frogs and fish are neither conscious nor unconscious, or that crocodiles do not sleep and later awaken. This is surely absurd, Sandy.

Sandy: So consciousness emerges phylogenetically with developed forms of sentience?

Adam: Quite so. Or, to be more accurate, perceptual consciousness does.

Sandy: And it has nothing to do with socialization?

Adam: No, of course not. There is something absurd about supposing that the young puppy or kitten, being licked by its mother, starts to develop a theory of its mother's mind. What would a puppy or kitten do with a conceptual framework if it came across one? It is nonsense to think that a newborn child is in the business of constructing theories, and it is nonsense on stilts to suppose that *the brain* should be in the model-construction business. Constructing theories and models is a highly sophisticated intellectual activity. It presupposes the mastery of extensive linguistic skills and possession of substantial understanding of the phenomena that one wishes to model and explain. And quite apart from that, what would a model of a mother and her brain look like, even if it were constructed by an adult, never mind by a neonate? Do you have a model of your wife and her brain? Haha!

Stranger: *Ja,* I think that is correct, Sandy. I also think that these confusions come from not having a clear conception of what consciousness is, and what forms it takes. It is futile, is it not, to ask what something is *for* if one does not yet grasp what the thing is. *So,* we must try to clarify some of these terrible confusions, otherwise our evening will not have been well spent – No?

Christopher: OK, so *you* tell us what consciousness is.

Stranger: No, no. I will try to get *you* to do so. Look, Chris, when you are hit hard on the head, – like *this* [*he gestures a blow on the head*] – you lose consciousness, no?

Christopher: Sure.

Stranger: And later on you regain consciousness, no?

Christopher: Yeah.

Stranger: And at night you fall asleep and are not conscious, and in the morning you wake up and are conscious again.

Christopher: Yeah, sure – but this is not what worries neuroscientists and psychologists.

Stranger: *Ja, Ja,* of course, but bear with me for the little while … Now, what is being conscious or awake *for*, what is its evolutionary advantage, uh?

Christopher: That sounds like a damned-fool question to me.

Stranger: *Ja,* of course it is. The serious question here is not what being awake or being conscious is for, but what sleeping is for.

Sandy: That is right – and that is actually a very difficult question for biologists to answer. We know what results from sleep deprivation, but we are very unclear why sleep is actually necessary, and even more unclear why some animals require so much more sleep than others.

Bruce: Right, but where is this getting us? I mean, you haven't even arrived at the hard question about consciousness.

Stranger: Bruce, we take it slowly, no? *Langsam, langsam, aber sicher* we say in German – slowly but certainly, *Ja.* Now, we speak also of conscious mental states, no?

Bruce: Sure.

Stranger: And a conscious mental state is not conscious, any more than a passionate belief is passionate, uh?

Bruce: I didn't get that?

Stranger: Well, a passionate belief is a belief that one cleaves to with passion, no?

Bruce: Oh yeah, I see what you mean. So what about conscious mental states – I can see that the mental state is not itself conscious – I mean, that doesn't make sense.

Stranger: That's correct. – A *conscious* mental state does not stand in contrast to an *unconscious* mental state, but to a *dispositional*, non-conscious, mental state; hmm. A conscious mental state is a mental state that we are in only *while we are conscious*, like feeling joyous or cheerful, or feeling depressed. But it does not go on when we fall asleep or are unconscious, hmm … But you may be cheerful by temperament, that is: have a disposition to feel cheerful when you are awake, just as you may be in a depression for months, that is: have a disposition to feel depressed when you are awake. Dispositional mental states persist over time, no matter whether you are conscious or not conscious.

Christopher: OK. But what about *unconscious* mental states?

Stranger: *Ja*, we can make room for that too. We must distinguish *non-conscious* mental states – that is: dispositional ones – from *unconscious* mental states. These are, as Freud made clear, mental states one is unwilling to acknowledge. But the story of Freudian, unconscious mental states is complicated. We need not to tell it now, huh.

So, Bruce, let us continue. What would you say is characteristic of animals that are conscious as opposed to asleep?

Bruce: Well, that's obvious enough. They are sensitive to visual and auditory stimuli, they see and hear things and respond to what they see and hear. They look around, watch things, listen to sounds, they sniff the air and smell things, and so on.

Stranger: Good … Hmm … Now, I hope you will grant me that you cannot intentionally and deliberately *become* conscious of something you see, and that you cannot be asked or commanded to *be* conscious of something in your field of perception.

Bruce: Well, I'd never thought of that! … Yeah, I can tell someone to look at something, but I guess I can't tell anyone to be conscious of something; or to become conscious of something. That's odd …

Stranger: No, no, Bruce, it is not at all odd. It only seems odd because it never occurred to you. Hmm.

Adam: And it never occurred to you because you didn't look around at the familiar linguistic data. After all, Bruce, you would never say to anyone: 'Become conscious of the scent of the roses!' Or 'Be conscious of the paintings!' – You know that perfectly well, but you don't call these linguistic facts to mind. Yet these are the very facts that will give you the insights you are looking for.

Stranger: *Ja, Ja.* – *So*, Bruce, you grant me that becoming and being perceptually conscious of something is not a voluntary act.

Bruce: Yeah, I guess that's right.

Stranger: *Ja.* And you will grant me that to be conscious of something implies knowing the thing to be present, just as to be conscious that something *ist so,* is to know that things are *so.* No?

Bruce: You mean I can't become and then be conscious of Adam if he isn't here at all, and [*he laughs*] I can't become conscious of the fact that he is a real pain in the neck if he isn't one? [*They all laugh, including Adam*]

Stranger: [*laughing*] *Ja.* Correct! Now what this shows is that to be perceptually conscious of something is a form of *cognitive receptivity.* Hmm! ... *cognitive receptivity.* It is not knowledge we *attain* by the use of the cognitive powers. It is not knowledge we *achieve,* but knowledge we *receive* – that is *given* us, thrust upon us when things in our perceptual field catch and hold our attention. Hmm!

Bruce: That's really interesting. Go on!

Stranger: Now reflect, hmm: what about all the things we perceive without having our attention caught and held by them? The things that we just see and hear, and the things we *deliberately* look at and listen to. Are we conscious of these things, uh?

Bruce: Well, ... I'm not sure what to say. If I say that I'm not conscious of all the things I see and hear, then it is going to sound as if I don't know what's in front of my eyes and that I am unaware of what's being said. So I'd prefer to say that I *am* conscious of everything I see and hear, since I do know what is in front of me and I know what's being said.

Stranger: But that would mean that you had your attention caught and held by something you perceived, whereas you did not. You *deliberately* watched what you were looking at, and you *intentionally* listened to the conversation going on, didn't you, uh?

Bruce: Yeah, I guess that's right too ... So what am I supposed to say?

Stranger: Well, now. Hmm ... You might just say that in these circumstances, the question does not arise, no? That it is neither correct that you were conscious of what you were intentionally attending to, nor that you were not conscious of what you were intentionally attending to. *Ja.* The question does not arise. The question of perceptual consciousness arises mainly when we are dealing with peripheral perception – when our attention is caught by something we glimpse on the periphery of our visual field, when we suddenly become conscious of the ticking of the clock in the background, or of the smell of dinner wafting in from the kitchen, uh?

Sandy: I see … Yes, that's very interesting. So the question of what consciousness is for disintegrated into the ridiculous question of what the evolutionary pay-off of being awake is. And now the question of what being perceptually consciousness of something is for, reduces to the question of what the capacity to achieve knowledge by peripheral perception is for. And that is just as trivial. No animal is going to survive for long if it cannot have its attention caught and held by things and events at the periphery of its sensory field. I mean, our ancestors would not have survived for long in the jungle without peripheral vision, would they? So the whole debate about the purpose, and the evolutionary warrant, of consciousness was just misconceived.

Adam: Quite so. It was no more than following a red herring.

Christopher: No, wait. I still don't get it. I mean, you have still not really explained what consciousness is. Y'know, in the USA we're getting this sorted out, and you haven't even mentioned what we've achieved.

Adam: Oh! And what is that?

Christopher: Well, the key to consciousness, and the heart of the mysteries of consciousness is the character of conscious experience. You didn't even touch on that, y'know. I mean, there is something it is like to see and to hear, to walk and to talk, to have experiences. It's obvious that there is nothing it is like to be a laser-jet printer or a brick wall, but there is something it is like to fall in love, to feel joyful or to feel depressed. And there is something it is like to be a man, or a woman. That's what consciousness is. Consciousness is the 'what-it's-likeness' of experience.[23] Conscious creatures have experience, and experience has a unique qualitative feel to it. For any conscious experience there is something it is like to have it, and for each conscious creature, there is something it is like to be it.

Stranger: Hmm. There are many kinds of things that we can be and become conscious of that we have not mentioned this evening. We can be and become conscious of facts and conditions, of information retained and born in mind. We can become conscious of what we are doing, either as agent or as observer. We can be self-conscious – in a number of different senses of the word. And so on, uh. But the one thing that we cannot be or become conscious of is what you are calling 'the what-it's-likeness of experience'. We must discuss this complicated business some other time. Perhaps when I am here on my next visit, uh? But I am afraid, my dear Chris, that I must stop now and retire to bed, as I have to catch the coach to Heathrow early tomorrow morning. [*He gets up to leave*]

Sandy: Well, thank you all so much, and especially you, sir [*he rises too and addresses the Stranger*], that was very instructive. I'll have to think about it, but you've surely put me on the right track.

Adam: Yes, many thanks, my old friend. That really was very helpful. Perhaps we can all meet again and take matters further on your next visit. Will you chaps turn the lights off, while I take our guest to his rooms?

[*Adam, his guest and Sandy leave the room, while Bruce and Christopher busy themselves turning off the lights and putting the glasses on the sideboard*]

Christopher: Y'know, I think I'm more bewildered now than I was before. I mean, surely there is something it is like to have conscious experience? And isn't there something it is like to be me? And there must be something it is like to be a cat or a bat, even if they can't tell us what it's like.

Bruce: Yeah … That's phenomenal consciousness, and it's got damn-all to do with peripheral attention or with being awake! I wonder what he's got against it.

Christopher: Right. I sure hope that guy comes back again and we can get Adam to bring him in. [*He turns off the last light*] OK, let's go. [*They leave*]

[*The room is dark, except for the flickering of the fire in the hearth. An eerie light falls on the bust of Socrates, illuminating it. Olympian laughter sounds in the Common Room for a while, and as it dies away, the light shining on the bust of Socrates fades*]

Notes

1 Those familiar with Plato's *Sophist* will recognize the affinity of the present stranger with Plato's. Their opening words more or less coincide.
2 T. D. Albright, T. M. Jessel, E. R. Kandel and M. I. Posner, 'Neural Science: A Century of Progress', review supplement to *Cell 100* (2000).
3 F. Crick and C. Koch, 'Mind and Brain', *Scientific American* 267 (1992), p. 111.
4 J. P. Frisby, *Seeing: Illusion, Brain, and Mind* (Oxford University Press, Oxford, 1980), p. 11.
5 D. Dennett, 'Consciousness', in R. L. Gregory (ed.), *The Oxford Companion to the Mind* (Oxford University Press, Oxford, 1987) p. 160.
6 D. J. Chalmers, *The Conscious Mind* (Oxford University Press, Oxford, 1996), p. 3.
7 T. H. Huxley, *Lessons in Elementary Psychology* (1866), p. 210.
8 J. Tyndall, *Fragments of Science* [1879] (D. Appleton, New York, 1891), 5th ed., p. 420.
9 I. Glyn, *An Anatomy of Thought* (Weidenfeld and Nicolson, London, 1999), p. 396.
10 J. Searle, *Brains and Science: The 1984 Reith Lectures* (BBC, London, 1984), p. 25.

11   C. McGinn, 'Could a Machine Be Conscious?' in C. Blakemore and S. Greenfield (eds), *Mindwaves* (Blackwell, Oxford, 1987), pp. 281, 285.

12   S. Greenfield, 'How Might the Brain Generate Consciousness' in S. Rose (ed.), *From Brains to Consciousness* (Penguin, Harmondsworth, 1998), p. 214.

13   M. Gazzaniga (ed.), *The New Cognitive Neuroscience*, 4th ed. (MIT Press, Boston, 1997), p. 1396.

14   N. Humphrey, 'The Inner Eye of Consciousness' in Blakemore and Greenfield (eds), *Mindwaves*, p. 379

15   A. Damasio, *The Feeling of What Happens* (Heinemann, London, 1999), p. 322.

16   The stranger's friend was the late Raymond Frey, a man of much wit and merriment.

17   Cp. Leibniz, 'The Principles of Nature and of Grace, based on Reason' [1714], §17 in P. P. Wiener, *Leibniz – Selections* (Scribner's, New York, 1951), p. 536.

18   S. Sutherland, *Dictionary of Psychology* (Macmillan, London, 1989).

19   P. Johnson-Laird, *The Computer and the Mind* (Fontana, London, 1988), p. 353.

20   H. Barlow, 'The Biological Role of Consciousness' in Blakemore and Greenfield (eds), *Mindwaves*, p. 361.

21   R. Penrose, *The Emperor's New Mind*, rev.ed. (Oxford University Press, Oxford 1999), p. 523.

22   N. Humphrey, *Consciousness Regained* (Oxford University Press, Oxford, 1984), pp. 35, 37, 380–81.

23   See T. Bayne, A. Cleeremans and P. Wilken, *The Oxford Companion to Consciousness* (Oxford University Press, Oxford, 2009).

SUPPLEMENTARY READING

M. R. Bennett and P. M. S. Hacker, *Philosophical Foundations of Neuroscience* (Blackwell, Oxford, 2003), chaps 9–12.

P. M. S. Hacker, *The Intellectual Powers: A Study of Human Nature* (Wiley-Blackwell, Oxford, 2013), chap. 1.

Norman Malcolm, 'Consciousness and Causality' in D. M. Armstrong and Norman Malcolm (eds), *Consciousness and Causality* (Blackwell, Oxford, 1984).

A. R. White, *Attention* (Blackwell, Oxford, 1964), chap. 4.

Fourth Dialogue

CONSCIOUSNESS AS EXPERIENCE – CONSCIOUSNESS AS LIFE ITSELF

Protagonists:

Christopher Cook: a distinguished American cognitive neuroscientist from California; middle-aged; pronounced American accent; dressed in jeans, sports jacket and open necked shirt. His views are a synthesis of a large number of different views among cognitive neuroscientists.

Jocelyn Thomas: The Stranger's companion; an educated lady in her late thirties; informally but well dressed; neutral educated accent. An inquisitive and intelligent mind, with a biological, rather than philosophical, background.

Bruce Palmer: an eminent Australian philosopher, a member of the self-styled 'consciousness studies community'; in his mid-thirties; relaxed; amiable; pronounced Australian accent; dressed in T-shirt and jeans. His views are a synthesis of many different views among philosophers who have nailed their colours to the masts of cognitive science.

The Stranger: the visitor from Vienna who participated in the previous dialogue on consciousness three weeks previously; late-middle-aged philosopher; dressed in a dark suit, with a bow tie; pronounced German accent coupled with a speech mannerism of punctuating his remarks with a slight 'Hmm' or interrogative 'huh?' A remote intellectual descendent of Plato's Eleatic stranger in *The Laws*. He is obviously familiar with the works of Wittgenstein.

Adam Blackstone: a middle-aged Oxford don; the host of the gathering; educated Oxford accent; dressed in a well-cut sports jacket and elegant tie, casual slacks; a linguistic philosopher.

The scene is a small Oxford Common Room. The room is panelled. Oil paintings of past fellows and of presidents hang from the picture rails. There is a bust of Socrates on a stand in the corner. Lighting comes from the central chandelier and

sconces on the walls. A fire is flickering in the fireplace. There is a well-stocked
bookcase against one wall. There are five comfortable leather armchairs arranged
around a low central table with books and papers scattered on it. On one side is a
sideboard on which there is an eighteenth-century table-clock as well as bottles of
drinks and glasses. The company is seated around the table with their drinks on it.

Christopher: OK. Look. Last time we met here, you two [*he gestures
towards Adam and the Stranger*] tried to persuade us that there is no great
mystery about consciousness, that it's just a biological phenomenon
that has as good an evolutionary explanation as any other biological
phenomenon. Now Bruce and I [*he glances at Bruce, who nods vigor-
ously*] have had time to think about this. You persuaded us that it's
misleading to talk of consciousness as a property of the brain – it's
a property of the living animal as a whole. OK. And y' showed us
that the question that bothered Sandy last time: the question of what
consciousness is *for* is a pretty dumb question. But we think that you
didn't put your finger on the heart of the matter. Right. Y' left out
what is really important about consciousness – y' left out the *qualitative
character of experience.*

Jocelyn: What is the 'qualitative character of experience'?

Bruce: Well, it's like this: despite what you two [*he nods towards Adam
and the Stranger*] explained to us at our last meeting, Chris and I are still
fascinated by consciousness. Maybe consciousness is not a mystery,
but it's still pretty mysterious. Look at it this way: we have sensory
inputs from the world around us – things touching our skin, molecules
of stuff coming into contact with our tongues and nasal passages,
sound waves impacting on our eardrums and light on our retinas.
And we have behavioural outputs: we respond behaviourally to these
stimuli. Now, somewhere between the input and the output, we have
conscious experience. Right!

Jocelyn: Go on!

Bruce: So the deep problem is to characterize the nature of conscious
experience. To do that is to characterize the subjective world – the
world of consciousness. And that's just what Adam and our guest
from Vienna did *not* do in our previous conversation.

Jocelyn: I see. [*She turns to the Stranger*] Is this world of consciousness
what Schopenhauer called 'the World as Idea' or 'the World as
Representation'?

Stranger: *Ja, ja.* There is a certain resemblance.

Adam: Quite so.

Bruce: I don't get this.

Adam: Look, old boy, if you start from inputs and outputs, you are bound to worry about whether there is anything in between except machinery. If what goes in is no more than radiation and the impact of particles, and what comes out is bare bodily behaviour, then of course it will seem as if we are witnessing *Hamlet* without the Prince. It will seem as if we have left out Life, Life with a capital L – Life as the *experienced world*. The World as Representation, as Schopenhauer said ... roughly.

Christopher: Yeah, that's right. That's why the idea of a zombie is so persuasive. Because a zombie has the same input and output as human beings, but without the intervening subjective experience. That's why we wanted to say that zombies are perfectly imaginable: they would have all the neural machinery, but without the conscious experience. They would respond to stimuli just as we do, but with no accompanying experience – without any inner life.

Adam: But, Chris, we showed you that the very idea of a zombie makes no sense!

Christopher: Sure; sure ... But it's still tempting.

Adam: Well, old chap, you must resist the temptation now that you know why it is the voice of sirens. And you must be careful in your talk of inputs and outputs. If the input is radiation and impact, then the output is neither behaviour nor action, but only *muscular contraction* and consequent *movement*, as well as *noises emitted*. If the output is behaviour in the full sense – that is, not bare bodily movement, but the myriad voluntary, intentional, deliberate acts and speech-acts we perform – then the input is not mere light waves and sound waves, it is what we see and hear, including what other people do and say to us.

Bruce: [*impatiently*] OK. OK ... Now, let's get back to the main point we didn't discuss. We can characterize the nature of experience – the nature of conscious experience – by reference to its specific qualitative character. The qualitative character of an experience is *what it's like to have it*.

Jocelyn: What do you mean 'What it's like to have it'? I can see that drinking this Courvoisier is very like drinking Remy Martin. But I can't see what that has to do with consciousness, Bruce.

Adam: Quite so. *That* doesn't. But that is not what Bruce means. He isn't asking for a comparison. One *can* ask 'What is Calvados like?' – that's a request for a comparison, to which a decent answer might be 'Rather like an apple schnapps'. One can also ask 'What is *drinking* Calvados like?', and that *could* be a request for a comparison – only this time a comparison of experiences, not of brands of liquor. To this too one

can answer in the same vein: 'It's rather like drinking apple schnapps'.
But that isn't what he means either. He doesn't want to know what the
experience *resembles*; he wants to know what it's like *for you.*

Bruce: Yeah. That's good. What we have in mind when we talk about
the qualitative character of experience is: what it's like *for the subject of
the experience to have that very experience.* Because that's what's distinctive
of experience: there is something it is like to have it. I mean, if you
take a sip of that Courvoisier, the experience has a very special character doesn't it? Go on, have a sip.

Adam: [*clutches his head and mutters to the Stranger*] Oh, no! This is like
running up debts to find out what a negative number is!

Jocelyn: [*taking a sip, while disregarding Adam's remark*] Yes, of course.
The cognac is very smooth, and there is a warm afterglow when one
swallows it.

Bruce: [*who hasn't noticed Adam's mutterings*] Yeah. Now, *every* experience has a qualitative character. For any experience you have, there is
something it's like to have it, and that fact is the key to consciousness.[1]

Jocelyn: I don't see why? Obviously one has to be conscious to enjoy
the cognac, but I can't see why that holds the key to consciousness.
I didn't know consciousness was locked up [*she giggles*]. One has to be
conscious to see one's finger nails [*she looks demonstratively at her finger
nails*] but I don't know that it's like anything in particular … except
perhaps like looking at one's toe nails [*she laughs*].

Christopher: [*a little annoyed by this frivolity*] Look now, Jocelyn … Is
there anything it is like for a brick to be a brick?

Jocelyn: No, of course not.

Christopher: And is there anything it is like for a laser-jet printer to
be a laser-jet printer?[2]

Jocelyn: Well, no; obviously not. I mean bricks and laser-jet printers
aren't alive.

Christopher: OK. But what about plants? Is there anything it is like
to be a cabbage? Or a cauliflower?

Jocelyn: Well, no.

Christopher: Why not? They're alive aren't they?

Jocelyn: Yes, but they're not conscious!

Bruce: Bull's eye! That's it. They're not conscious. But you are. And
there *is* something it is like to be a human being, isn't there?

Jocelyn: Er … Yes, I suppose so.

Bruce: Well *of course* there is. Now, there isn't anything it is like to *be*
a glass of cognac. But there is something it is like *to drink* a glass of
cognac – to have a conscious experience of drinking it.

Christopher: Yeah. And an experience is a conscious experience only if there is something that it is like for the subject to have it. Now [*turning triumphantly to Adam and the Stranger*] that's the essence of consciousness. To be conscious is to have conscious experience. And there being something it is like *for one* to have an experience – *that's* the qualitative character we're talking about.

Bruce: There's a further point. It's not just that conscious experience has a qualitative feel to it. And it's not just that it is essentially subjective – that it exists only as *someone's* experience – but it essentially involves a *subjective point of view*. A conscious creature is a creature that has a subjective point of view – an experiential perspective.[3]

Stranger: Excuse me, but what is a 'subjective point of view'? Or an 'experiential perspective'? Hmm.

Bruce: Each of us experiences the world from a point of view. Right. Every conscious creature perceives the world from *its* distinctive point of view. You can know roughly what it's like to experience things from the point of view of another human being, but you can't really know what it's like to experience the world from the point of view of a dog, a cat or a bat. I mean, you can't really know what it feels like to echo-locate, or to smell the world – to experience the world – through the nostrils of a dog. But y' know that bats and dogs are conscious creatures, don't you? I mean, you're not a Cartesian who thinks that animals aren't conscious beings at all? Y' know that other animals have conscious experiences. But y' don't know what it's like to be, say, a dog or a bat. Now, for any kind of conscious creature, there is something it is like to be a creature of that kind. There is something it is like to be a bat, and for sure there is something it is like to be human.

Christopher: Yeah, you two [*he glances at Adam and the Stranger*] didn't talk about any of this when we last met. But this is the cutting edge of cognitive science today. We're trying to get clear about the what-it's-likeness-of experience,[4] about the essential nature of consciousness. And what we neuroscientists are trying to do is to identify the NCC which …

Jocelyn: What is the NCC?

Christopher: The NCC is the neural correlate of consciousness. What we want to find out is what neural processes are correlated with conscious experience. We want to know why it's like what it's like to see red, and why we have the experience of redness when light of around 700 nanometers wavelength impacts our retina. And we want to know why there isn't anything that it's like for a computer equipped with a camera to detect exactly the same information.[5] There is something it

is like to see red, and something it is like to see green. That's the qualitative character of experience – the phenomenal character of experience. And what we want to understand is what the neural correlate of
conscious experience is. Look, qualia – to use a good American word
introduced by C. I. Lewis[6] – qualia are *experiential feels*, like the pain
of a stubbed toe, or the red that you experience when you see a ripe
tomato, or the taste of a pineapple. These are the qualities of subjective experience. I mean, the subjective taste of a mango, say, can
be known or understood only from the perspective of those who have
had the experience.

Stranger: Well, my friends, I have been listening to you carefully, but
I am finding it very difficult to follow. Hmm. I think that you must
slow down; not take things so quickly. After all, we are not in a hurry
are we, huh? We must take things very slowly.

Adam: Be a little more angelic!

Stranger: What do you mean 'angelic', Adam?

Jocelyn: [*sotto voce*] He means that fools rush in where angels fear
to tread.

Stranger: Ah! *Ja*. I see … Alexander Pope, no? Very amusing … haha!
So, yes, we must not rush in. Hmm. It seems to me that there are many
different threads here, which we must unpick, yes.

Bruce: What d'you mean – many different threads?

Stranger: Well, you surely have noticed that you and Christopher have
very different notions of the 'qualitative character of experience', huh?
What *he* means is the sensible qualities of objects of experience as we
sense them – the redness of roses, the warmth of the fire, the sound of
the trumpet, *Ja*. And what *you* mean is the qualitative character of the
experiences themselves, huh! Not of the objects of experience. These
actually are quite different, and they need to be kept separate.

Christopher: Now, wait a moment. I don't see the difference here.
Surely there is something it is like to see the redness of a rose. We want
to understand the redness of red. I mean, the redness that you perceive so vividly can't be precisely communicated to another human
being.[7] But it may be possible to discover the *neural correlate* of your
seeing red. Then it may be possible to explain why you experience the
vivid sensation of red, and why some other neural correlate makes
y' see blue. Now if it turns out that the neural correlate of red is
the same in your brain as in mine, then it will be scientifically plausible to infer that your experience of redness is qualitatively the same
as mine.[8] That would solve a problem that philosophers have been
grappling with for centuries.

Jocelyn: You mean how I can know whether you perceive the world the same way as I do? Whether your experiences of colour are just like mine or quite different? The inverted spectrum and all that?[9]

Christopher: Yeah. That's right. Neuroscience will be able to solve problems that philosophers have struggled in vain to solve. Sitting in armchairs is no way to go about solving serious problems. We need the right experimental data and a good theory.

Adam: [*annoyed*] Chris, you can no more solve a conceptual problem in a laboratory or observatory, than you can prove a mathematical theorem by taking astronomical observations or doing physical experiments.

Stranger: Adam, I am in agreement with you, of course. But let us not go down that way now, hmm. We must keep to our theme. For although we cannot solve all of the problems 'at one fell sweep', as you say, we can perhaps solve a selection of the problems this evening. *So* let us concentrate on the qualitative character of conscious experience, huh?

Adam: 'At one fell swoop'; Yes, all right. But some evening we must clarify to the scientists that while we can solve some of their problems, they can't solve *any* of ours.

Christopher: [*indignantly*] Isn't that just a little bit arrogant, Adam?

Bruce: [*indignantly*] Hey, what d'you mean?

Stranger: [*waving his hands to calm them down*] Now, now … never mind. Be still. Let us concentrate on the qualitative character of experience, huh? We must let some light into this darkness, huh?

Good. Hmm. Now. We must clarify what exactly you mean by the qualitative character of experience, huh. Do you have in mind the fact that we can all see colours, hear sounds, taste tastes, smell smells and feel warmth and cold, and other tactile qualities? Or do you mean that every experience, no matter what it is, whether it is perceiving, or feeling an emotion or mood, or even just thinking about something – which is not an experience at all – has a special qualitative character that uniquely identifies it as the kind of experience it is?

Then we must make clear what exactly you mean by this curious phrase 'there is something that it is like to …'

Bruce: I don't see why you say it's curious, it's …

Stranger: Excuse me, Bruce, please let me finish putting the problems on the table, Hmm.

Bruce: Oh, I'm sorry.

Stranger: *Ja*, this phrase 'there is something it is like' is more problematic than you think. So we must examine it carefully. We must see how the question 'What was it like for you to experience this or that?' is actually used, and what kinds of answers it can be given; uh. Then we must look carefully at the question 'What is it like to be this or that?'. Hmm. We must also make clear some things about 'subjectivity' and a 'point of view' – for these are, as Adam said, the songs of sirens to lure us onto the rocks. Good. Now let us first turn to the qualitative character of experience, huh. You are inclined to say that whenever you have a perceptual or sensory experience, then there is something it is like for you to have that experience, yes?

{ **Bruce:** Sure.
{ **Christopher:** Yeah, that's right.

Stranger: And what of you, Madame Jocelyn. Do you think that there is something it is like to have an experience?

Jocelyn: Well, I'm not sure. There is something it is like to have a headache, isn't there? It's unpleasant, isn't it? And one feels like shutting one's eyes, or lying down in a dark room. And I agree that all sorts of experiences are very agreeable, like swimming in the sea or lake in the summer, or mountain walking in the Lake District. It's invigorating, the landscape is beautiful and the air is clean … On the other hand, I'm not sure [*she demonstratively looks at her fingernails*] that looking at my nails is particularly interesting or enjoyable? [*She giggles*] And seeing the lamp posts as one walks down the street isn't interesting, but it isn't boring either. And although not being able to breathe is terrible, breathing in normal circumstances isn't anything in particular.

Stranger: *Ja; Gut.* So at first look, it seems that if we are asked whether there is anything it is like to do or undergo something, we can answer 'Yes', if it is agreeable, and 'No', if it is disagreeable. Huh? And if it is neither the one nor the other, then we should reject the question. If you asked me what it is like to look at the buttons on my shirt-front, I should ask you what you meant, no?

Adam: So it is just false that for every experience, there is something that it is like to have it. Some experiences are such that there isn't anything it is like to have them.

Christopher: No, I don't agree. I mean, whenever y' see anything y' have the visual experience of colors. Right. And there is something it is like to see red, something that's quite different from what it's like to see green. And when y' hear the clock ticking, there is something it is like to hear it, something quite different from what it is like to hear the clock chime. And as my friend John Searle has pointed out, there is something that it is like to think that two plus two equals four. It feels quite different from thinking '*zwei und zwei sind vier*'.[10] And David

Chalmers told us that when he thinks of lions, there is a leonine whiff about the experience.[11]

Adam: You're joking! I hope he doesn't think of skunks too often!

Stranger: Adam, shush! What matters is this: to ask *what it is like* to do something is not the same as to ask *what one associates* with doing something; huh? Perceiving roses one may *associate* with *Der Rosenkavalier* – which one may not like at all – but smelling roses is most agreeable, no? What one associates with an experience is not the same as what the experience is like. Still, allow me to delay the examination of perceiving sensible qualities for a moment. I would like first to deal with the qualities of experiences. Hmm!

Christopher: OK. Sure. Go ahead.

Stranger: Thank you. *Gut.* Now, Fraulein Jocelyn correctly distinguished between experiences that can be said to have a certain qualitative character and experiences that are, so to say, neutral. The former can be characterized as having a certain hedonic or, hmm, can I say 'anti-hedonic'? *Ja* – anti-hedonic character, ah. But other experiences are colourless, that is: hedonically neutral. Which is perhaps just as well, huh? For if it were not so, we should be swamped with pleasantnesses and unpleasantnesses, no?

Bruce: But look! Every experience *must* have a unique *qualitative feel* to it. Its qualitative character is precisely what makes it the experience it is.

Stranger: Do you mean, perhaps, that what makes one pain different from another is that one is a stabbing pain and the other is a burning pain, huh?

Bruce: Yeah, that kind of thing. What differentiates different pain experiences is their phenomenal character.

Stranger: Hmm. Well, we differentiate pains, which are sensations, by reference to their phenomenal characteristics. *Ja.* Hmm. Now what differentiates seeing Adam from seeing Christopher, huh?

Bruce: Well they don't look at all alike. I mean Adam wears glasses and Chris doesn't, Adam has dark hair and Chris's hair is sandy, and so on.

Stranger: Precisely. You don't differentiate your seeing what you see by reference to what it is like to see the things you see – whether it is delightful or horrible, charming or dull, ah – but by reference to *what* you see. You may be delighted to see your friends, but the identity of the seeing, so to say, is given by what is seen. What it was like for you to see what you saw is a further question, which may or may not have an answer. In the case of my buttons [*he chuckles*], it does not, huh?

Jocelyn: You mean that sensations have phenomenal qualities, but perceptual experiences don't?

Stranger: *Nein, nein.* Didn't you say that seeing the views in the Lake District is wonderful, life-enhancing, huh? No, no, the point is that whether or not there is a phenomenal characteristic to perceiving something, or to perceiving that something is so, does not determine the identity of the perceiving, *so* to say. The lack of any hedonic or anti-hedonic character does not imply the absence of any experience, that is: of seeing or hearing, feeling or smelling, huh. But the story does not end there.

Adam: Why not? You've shown that it is not, *in general*, a mark of experience – of 'conscious experience', as they call it – to have a quali- tative character. Why doesn't the story end there?

Stranger: Because we have not dug down to the roots of the matter. Hmm. [*He chuckles*] We must dig deep if we want to uproot the illusions of Reason, ah. Now, Fraulein Jocelyn, you said, quite correctly, that some experiences are agreeable and some are disagreeable. *So*, when you *characterize* an experience you have, to what question might you be replying, huh?

Jocelyn: I'm not quite sure what you mean ... I suppose that I should say that something was pleasant or unpleasant, exciting or boring, if I were asked what it was like to do it.

Stranger: Precisely. You might be asked, 'What was it like to see moun- tain gorillas in Rwanda?' or 'What was it like to fly in a balloon over the plains of the Serengeti?', huh. Now, Bruce, how do we get from this to the idea that *there is something it is like* to have an experience?

Bruce: How d'you mean?

Stranger: Well, think now, ah! Jocelyn is asked, 'What was it like?' and she replies, 'It was wonderful'. So there was someone, namely Jocelyn, who found it wonderful to have this experience. And now ...?

Bruce: Well, there was something, namely an experience that Jocelyn had, that was wonderful.

Stranger: Precisely. And now what was the characteristic feature of Jocelyn's experience?

Bruce: Well, I just said what it was – it was bloody marvellous.

Stranger: So there was something that Jocelyn's experience was, ah? Namely: wonderful. Hmm?

Bruce: I don't get it?

Stranger: Well, what has happened to the 'like', huh?

Bruce: You've lost me.

Adam: Don't you see! The existential generalization of the answer to a question of the form 'What is it like to V?' is not 'There is something *it is like* to V', but 'There is something it *is* to V'.

Christopher: I don't understand that. What is existential generalization?

Stranger: Never mind about existential generalization. If you are asked, 'What was it like for you to visit Florence for the first time?', then you will say, 'It was wonderful', no?

Jocelyn: Yes, of course.

Stranger: So *there was something it was* for you to visit Florence, namely wonderful. Hmm. Not '*It was like* wonderful'.

Adam: Not unless you come from California.

Stranger: Haha! *Ja*, just *so*. There was something *it was* for you to visit Florence, namely wonderful, *Ja* – not: 'there was something it was for you to visit Florence, namely *like wonderful*'; and not 'There was something *it was like* for you to visit Florence' either, huh?

Jocelyn: [*trying it out*] What was it like for you to visit Florence? – It was wonderful. – So there was something that it was for you to visit Florence, namely wonderful. – Yes, that's right. The 'like' just disappears. Why didn't I notice that?

Adam: That's what I meant when I said that the existential generalization of a question concerning the qualitative character of an experience excludes any 'like'-s. The question 'What was it like for you to …?' asks for the *characterization* of the experience. One could just as well ask 'What sort of experience was it?', And the answer specifies just that: it was a wonderful experience; or: it was a horrible, terrifying one; and so forth. But not: it was a sort of wonderful one – or a sort of horrible one!!

Jocelyn: So actually there isn't *something it is like* to have an experience at all?

Stranger: Precisely so, Jocelyn. Hmm. For some experiences there may be something it *is* to enjoy or suffer them. For other, neutral, experiences there is not, huh. But there is not *something it is like* to have experiences at all. That is just an illusion – a *grammatical* illusion. [*He turns to Bruce and Christopher*] You see?

Adam: Yes. And it is a grammatical illusion that stems in part from our relative unfamiliarity with operations of second-level quantification over properties.

Stranger: That doesn't really help anyone except those familiar with the predicate calculus, hmm. We here can disregard that. It is enough to grasp that the roots of the confusion are grammatical, and to see what features of grammar are being twisted.

Adam: Absolutely. And it's absolutely time for a top-up. [*He rises from his chair and goes over to the sideboard with the drinks*] Jocelyn, another Courvoisier?

Jocelyn: Yes, a small one please. [*Adam tops her glass up*]

Adam: [*turning to the Stranger*] My friend, another cognac, or something else?

Stranger: No, another cognac would be very welcome, thank you. [*Adam pours him a cognac*]

Adam: Chris?

Christopher: Oh, thanks, Adam. I'll help myself. [*He rises, pours himself a fresh drink and resumes his seat*]

Adam: Bruce, what about you?

Bruce: Thanks, Adam. Another Scotch please – and don't be too modest, haha!! [*Adam fills Bruce's glass up, then replenishes his own and sits down*]

Adam: All right. Now [*he turns to Bruce and Christopher*], do you see why this hysterical university talk about the 'what-it's-likeness-of-experience' is just empty chatter?

Christopher: Well, I think so. I'll have to digest it all. But even if you're right about 'what it's like' to have this or that experience, that it's pleasant or unpleasant, and all that, it still seems to me that seeing red has *a very distinctive quality*, altogether different from seeing green. Right! That's what *I* had in mind when I spoke of the *qualitative character of experience*. I mean, we experience the redness of red, and the greenness of green, don't we.

Stranger: Well, my friend, do you mean that you want us to examine the question of what seeing red is like? Or hearing a musical note, say A-sharp? Or tasting sweetness or bitterness?

Christopher: Yeah. That's right. Because we all know what it's like, but we can't put it into words. That's what *I* mean by 'the qualitative character of experience'. A sensation can't be directly conveyed to someone else, can it? I mean, I can't give you my experience to scrutinize, can I? So how can I explain my experience of seeing red to you?[12] That's what's mysterious about consciousness. It's not communicable. Or at any rate, not fully communicable.

Stranger: Yes, I see. Very interesting, Christopher. Hmm. Now you have, without noticing it, moved onto a quite different track. For you don't mean 'What was such-and-such an experience like – was it agreeable or disagreeable?' For as we have just seen, most seeings and hearings are, so to say, colourless – hedonically neutral, *Ja*. Seeing the buttons of your shirt is neither agreeable nor disagreeable, hmm. And if someone were to ask me, 'What is it like to see the colour of Christopher's buttons?', I should reply 'What *do* you mean?', hmm. So what *do* you have in mind when you ask 'What is seeing red like?'?

Do you mean 'What does seeing red *resemble*?' or do you mean 'What *is it* to see red?' Huh?

Christopher: I don't see the difference. How can I explain what it is to see red except by explaining what it's like to see red? And how can I explain *to another person* what it's like *for me* to see red? After all, only *I* know what it's like for me to see red. I mean, *I can't give you my experience.* And *I* can't really know what it's like *for you* to see red.

Stranger: Very interesting. And why can't you know this, huh?

Christopher: Well, in order for me to know what it's like for you to see red, I should have to get inside your head, wouldn't I.

Adam: Is that because that's where you would see it?

Christopher: Yeah, sure. I mean, that's where you see what you see. And if I could see what you see, I'd have to get inside your head.

Adam: I see. Now [*he looks around*] there are three red things in this room. Where are they?

Christopher: [*suspiciously*] What d'you mean?

Adam: I just mean: where are the three red things in the room? [*He rather demonstratively fingers his red and orange tie*]

Christopher: [*warily*] OK, so your tie is red, the ruby in Jocelyn's ring is red and there is some red in the background of that painting [*he waves his hand towards one of the portraits on the wall*]

Adam: So do you mean that my tie is round my neck, but its colour is in your head, old boy? Or that Jocelyn's ring is on her finger, but the colour of the ruby is in your mind?

Christopher: I ... I 'm not sure. I'm not sure what to say.

Adam: Well let me help you. If the colour of the ruby were in your head, then the ruby would look like a diamond. And if the redness of my tie were in your head, then it wouldn't *be* the redness of my tie, and my tie would be colourless and transparent! In which case I'd return it to the shop. Worse still, you wouldn't be able to see the colour of my tie, since you can't see anything inside your head, unless it is opened up – and even then you'd need a mirror.

Christopher: Oh! I see! ... Or actually, I don't. Now I'm completely confused!

Stranger: We must take things slowly, Christopher. There are ancient confusions about perceptual qualities here, which are a subject for some other occasion.[13] We need not enter into the treacherous sands of secondary qualities now. But we must look at some of the unclarities about ostensive definition of names of perceptual qualities, such as colour names.

Christopher: I don't follow you. What's ostensive definition?

Stranger: Well, Christopher, how would you answer the questions 'What is red?', or 'What is violet?' Or even better, questions about shades of colour, such as 'What is maroon?' or 'What is magenta?'

Christopher: Y'mean to a child?

Stranger: *Ja, ja.* To a child; or to a foreigner.

Christopher: OK. I guess I'd point at Adam's tie, and say 'That's red' or at the background of that painting and say 'That's maroon'.

Stranger: Good. Hmm. Now, when you give this ostensive, that is: *pointing* definition – definition by pointing – do you mean 'That tie is red' or 'That colour is red'?

Christopher: I'm not sure. What's the difference?

[*Silence for a few seconds*]

Jocelyn: Oh, I see. You can't say 'That tie is red', because that's a true proposition, and one can understand it only if one already knows what the word 'red' means.[14]

Christopher: Yeah! But I do know what the word 'red' means.

Jocelyn: Of course you do, otherwise the questioner wouldn't be asking you. But *she* doesn't know what it means.

Christopher: OK. So what should I say?

Jocelyn: Well, when you point at the sample you have selected, you should say 'That *colour* is red' or 'That *colour* is maroon; or magenta, or whatever'. That explains what the colour words mean.

Bruce: Hey, slow down. Now you're losing me. How can he understand what the word means unless he has the experience of seeing red. I mean, red is *the content of his subjective experience*, isn't it?

Jocelyn: Well, I … [*she looks at the Stranger*], … I'm not sure which way to go now.

Stranger: You are doing very well, Jocelyn, hmm. Very well. But let me see if I can help a little. Christopher said, surely correctly, that to explain what a colour word means to a child or foreigner, we would naturally point at something red and say 'That colour is red'. That is what is called 'an ostensive definition', hmm. We don't point at or into our head or his head. We point at some object that is visible to us both, *Ja*? We use this object as a sample, or standard, for the correct use of the colour word, and we give a rule for the use of the word.

Jocelyn: Yes, I see. We aren't *using* the object we're pointing at as an *example*, but as a *sample*. We're *explaining what the word means*, not giving examples of things that happen to be red.

Adam: That's right. A sample is like a pattern or prototype. It's an object for comparison or for copying. It is a standard that objects have to meet to qualify *as* such-and-such things, or things with such-and-such qualities. So we have samples of Meissen porcelain, of Worcester plates or of Copenhagen-blue cups and saucers. Or samples – swatches – of different kinds and colours of silks and materials or of carpets. And so too we have samples of colours – on colour charts – and samples of length – such as rulers and tape measures, and we have samples of sounds and tastes too.

Stranger: *Ja, ja.* Good. We *give* the meaning of the colour word by pointing at the right colour sample, but, of course, *what we point at* is not the meaning of the word.

Bruce: Slow down. Why not?

Stranger: Well, among many things, if what we point at when we explain what a word means, for example the word 'chair' or the word 'football', were the meaning of the word, then you could sit on some meanings and kick others [*he chuckles*]. No, no – the sample we point at when we explain what the colour word 'red' means is not the meaning of the word 'red'. Hmm. But what is important, Bruce, is that the sample is a public object, visible to all. To learn what the word 'magenta', for example, means, is to learn *to use* the sample pointed at as a standard for the correct use of the word, huh? For the explanation gives us a *rule* for the use of the word explained, namely: anything which is *that* ☞ [*he points*] colour is correctly described as being magenta. Hmm.

Bruce: But surely, the guy who's learning has to have the experience of seeing magenta.

Stranger: That can be misleading, huh? Of course, he has to be able to *see* magenta, otherwise he would not be able to identify things that are magenta, or to use the sample of magenta *as* a sample. The sample is a *paradigm* by reference to which he can both *apply* the word 'magenta' to things of that colour, and *use* as a court of appeal by reference to which to justify his application of the word.

Bruce: Yeah, that's what I meant. What's crucial is *the experience he has* when he looks at it.

Adam: No, no. What is crucial is his perceptual ability and its exercise. He doesn't *use his experience* as a sample or paradigm by reference to which to apply or justify the application of the word 'magenta' to an object. The sample is the *object* used as a standard or warrant for correct application.

Bruce: No, I disagree. Surely what we're talking about is the *content of his visual experience* when he looks at the object we point at.

Stranger: Slow down, Bruce. If you try to jump every gate, you will fall. You must *open* them, not *jump* them! Hmm. Now, 'visual experience' is just *seeing*, isn't it?

Bruce: Yeah, I guess so.

Stranger: And the phrase 'the content of visual experience' just means 'what you see', no?

Bruce: OK.

Stranger: *So* 'the content of what you see when you see the colour of something red' means the same as 'what you see when you see the colour of something red', no?

Bruce: OK.

Stranger: And what do you see when you see the colour of something red?

Bruce: Oh ... Well, er ... I see red.

Adam: Yes – so we can cut the cackle about 'the content of the visual experience you have when you see something red', and just stick to 'seeing red'.

Stranger: Exactly *so* ... *So* we can explain what colour words mean to anyone who has normal colour vision – who has the ability to discriminate between colours as normal human beings can, *Ja*. It is not his *seeing red* that explains what the word 'red' means. It is *the rule for the use of the word 'red'* that explains what it means. And the rule for the use of the word 'red' is that anything that is *that* ☞ [*he points at the picture*] colour is correctly described as being red. There is no mystery here, huh? And there is nothing – how do you say – ineffable.

Christopher: But we can't explain what it is like to see red. I mean, we can't say what it's like. That's one reason we want to discover the NCC – the neural correlate of the consciousness of seeing red.

Jocelyn: Now we're back where we started!

Stranger: *Nein!* Not really. We have climbed much higher, and if we look down without losing our heads, we shall be able to see where one usually loses one's footing, hmm.

Now, we must be careful what exactly we have in mind when we say that we cannot explain what it is like to see red. It is now clear, I hope, Christopher, that we are no longer concerned with whether it is agreeable or disagreeable to see the red colour of an apple or of some red paint, of blood or of a Chateau Rothschild, huh?

Christopher: Yeah, that's irrelevant now.

Stranger: Good. So are you asking what red resembles, huh?

Christopher: I'm not sure.

Stranger: Well, there does not seem to be anything difficult about that question, huh? One can say that red resembles pink, only with much less white, huh?

Christopher: Yeah, but that wasn't what I meant. I want to say that red is indescribable. That you can't describe what red is, you have to experience it. And each person has his own experience of seeing red. It's something completely subjective. That's why it's incommunicable. I mean, I can't give anyone else my experience, can I?

Adam: Oh, oh! Here we go.

Jocelyn: No; ... I see what Chris means. You can't describe the difference between say red and green. You have to see them – then you know the difference between them.

Christopher: Yeah, that's right. You can't describe the difference between red and green. But maybe, when we have found the NCC, we'll be able to say something about the difference, and maybe we'll be able to settle whether we all really see the same colors or not.

Stranger: Very interesting. I shall not examine the question of whether we see the same colours, the so-called inverted spectrum problem. It is getting late and we still have enough on our plate. Hmm! But we must look at the idea that we cannot describe the difference between red and green, but only see that they are different.

Christopher: Yeah. I mean, red is red and green is green. The only difference between them is that they are different! That's all that one can say.[15]

Stranger: Well, we can say that red is the colour of tomatoes and green is the colour of grass. Isn't that a difference, hmm?

Christopher: No, that's just a coincidental difference, not an essential one. I mean, grass could have been red and ripe tomatoes might have been green. But that's not the *real* difference. You can *see* the real difference, but you can't describe it.

Stranger: Very good. Now, can you describe the difference between a chair and a stool, huh?

Christopher: Sure. A stool doesn't have a back to lean against, but a chair does.

Stranger: *Ja.* And now can you describe the difference between a tricycle and a bicycle?

Christopher: Yeah, sure. A tricycle has three wheels and a bicycle has only two. I don't see what you're driving at.

Stranger: Be patient, Christopher, hmm. Now, how do you describe differences between such things?

Christopher: What do you mean?

Stranger: Well, I asked you to describe the respective differences between each of two pairs of things of a certain kind, no. In the first case, between two kinds of furnitures, and in the second case, between two kinds of cycles.

Christopher: Yeah. And I described the differences between them.

Stranger: By specifying different properties that distinguished them, no.

Christopher: Yeah, sure. How else can one specify differences between two things?

Stranger: *Ja, Ja*; one way to describe the difference between two kinds of things is to specify a feature or property that the one has, and that the other does not, no?

Christopher: [*baffled*] OK. So?

Stranger: *So*, when you are asked to describe the difference between two *properties*, like red and green, you obviously cannot do so on the pattern of describing the differences between two kinds of *objects or substances* by specifying their differentiating properties. So whether you can be said to describe the difference between two properties simply depends on what you are willing to call 'a description' and, in particular, what you are willing to call 'a description of the difference between two properties', huh?

Christopher: I don't follow you.

Jocelyn: It's clear enough, Chris. Look, you describe the difference between two substances by specifying a differentiating property that one substance has and the other lacks. But when you're asked to differentiate red and green, you can't specify a differentiating property that one substance has and the other lacks, since red and green aren't substances but properties. And you can't describe the differences between the properties by mentioning their differentiating properties, since properties don't have properties.

Christopher: Yeah, I thought that was what I said.

Jocelyn: No, Chris, you said that you can't describe the difference. But you *can* describe the difference – only not on the model of differentiating between two substances by reference to a differentiating property. You can say that green is the colour of grass and red the colour of tomatoes. Here you differentiate the two properties by reference to different substances that have them. But you rejected that, because it is a mere matter of fact that grass is green and tomatoes are red. And you can describe the difference between red and green by saying that red is what pink becomes when you remove the white, and green is … oh, well, green is the complementary colour to red. But you rejected that too, since it is not the same as differentiating two kinds of substance by reference to a property one has which the other lacks. So whether you can describe the difference between two colours, or two perceptual properties, simply turns on what you are willing to *call* 'a description'. And when you say that one *can't* describe the difference, all that means is that you aren't going to accept anything as a description of the difference.

Adam: Exactly, and that is not a *metaphysical* limitation, but a constraint you want to place upon the use of the phrase 'describing the difference'. You want to limit it to describing the differences between substances; and you want to reject the analogy between describing the differences between substances and describing the differences between properties.

[*Silence for a few moments*]

Bruce: OK. I think I can see what you're driving at. But all that takes us a long way from consciousness!

Adam: Well, not as far as you may think, Bruce. For one thing, it has become clear that whatever the neural correlate of seeing red is going to turn out to be, it is not going to explain what it is to see something red, nor what it's like to see something red. Nor is it going to resolve the puzzle about whether what you see to be red is the same as what I see to be red.

Stranger: [*hastily (looking at his watch)*] But we are not going down that road now, Adam! Ah.

Bruce: Right. But what about the subjectivity of experience? I mean, subjectivity is the ontological mode of experience. It involves a subjective point of view. And that's what makes conscious experience what it is.

Stranger: Hmm. I doubt whether ontological modes get us any further forward, they just lift our feet off the firm ground, hmm. Of course, every experience is the experience of some sentient being, *ja*; just as every act is the act of some agent. There are not songs without singers, and not dances without dancers.

Jocelyn: But that isn't what people mean when they say that experience is subjective. There are no lengths without things that are so-and-so long, and no heights without things that are so-and-so high, but length and height aren't subjective.

Bruce: Yeah, that's right. I mean, conscious experiences exist only from a first-person point of view.[16] When I know about your conscious experiences, I have knowledge that is quite different from the kind of knowledge I have of my own experiences.

Stranger: *Ja*, that is a very interesting conception. The whole business of so-called privileged access to one's own experiences, and the alleged knowledge of one's own current experiences by means of introspection lies at the root of the whole European tradition of epistemology. But to dig this up at this time of the evening is, I am afraid, too much. So I will make only one remark about points of view. Hmm. The whole idea of a point of view is muddled, huh.

Jocelyn: Why do you say that? It is a perfectly ordinary phrase with a perfectly respectable use.

Stranger: *Ja, ja.* Of course. But it is being misused by philosophers. Reflect: a point of view, originally, meant a point in space from which something is viewed, yes. Then it became stretched to mean a set of considerations that make up the framework within which one considers a question, hmm. So, we have the moral point of view, and the political or economic point of view, or the military point of view. And we have the idea of the point of view of some interest group, as when we speak of the European or American point of view, hmm; then we consider the matter from the perspective, so to say, of the interests of that group, *Ja*. Now, it is in this sense that I can speak of my point of view. This may mean, taking into consideration my interests, or taking as a starting point my opinions. Hmm.

Bruce: But that's not what I mean. What I mean is that every one sees the world from his point of view.

Stranger: Precisely. And there is no such thing. For all this amounts to now is your adamant insistence that you are a subject of experience. And no one wants to deny that, huh? Of course, you literally see what you see from a *point de vue*, yes … *a point de vue* that you occupy. But when you move aside, someone else can see what you saw from the very same point, no?

Bruce: But he can't have my experience!

Stranger: So it seems. But the chimera of private ownership of experience must be vanquished some other time. It is a remarkably tenacious little problem, uh. But not now.[17] Let us finish off what we have to say about consciousness.

Bruce: OK. Let's go back to square one. It seems to me that you haven't made clear what you take to be a conscious experience. If a conscious experience is not an experience which is such that there is something it is like to have it, and if most conscious experiences are, as you put it, hedonically neutral, then what the hell *is* a conscious experience?

Adam: Well, if you hadn't started out with inputs and outputs, and if you hadn't slipped on the banana skin of zombies, then you would have realized the simple truth that all talk of conscious *experience* is pleonastic. In your terms, *all* experience is conscious experience. Or, more lucidly, one only enjoys or undergoes experiences while one is conscious. When you are unconscious, or fast asleep, you don't have *any* experiences. Intransitive consciousness, that is, being awake as opposed to asleep or conscious as opposed to unconscious, is a condition of the possibility of experiences. There is no such thing as unconscious or non-conscious experience.

Bruce: But what about dreams? Surely to dream is to experience all manner of things.

Adam: No, no, Bruce – dreaming is a fresh can of worms, which we can open some other evening. But to dream is not to experience anything, although you often dream that you experience things. To dream that you experience something is no more to experience something than to dream that you die is to die. Dreaming is not a form of experience at all, even though when you wake up from a vivid dream it may seem to you for a few moments that you have just experienced various things. To talk of conscious experience is just to talk of experience which, tautologically, is enjoyed or undergone while one is conscious.

Christopher: OK. You've explained why it's an illusion to think that there is something it is like to have a conscious experience. Rather, some experiences are pleasant and so forth; and others are unpleasant or disgusting and so forth; and most are neutral. OK, I accept that. And you've explained that it's an illusion to think that the qualities of perceptual experiences, I mean, the redness of the red we see, or smelling violets or roses, and so forth, are indescribable. OK, I think I can go along with that. And now you've given what seems to me to be a trivial explanation of what conscious experience is. Now, I'm *not* sure about that. It seems to me that a conscious experience is an experience of which one's conscious. It's an experience you're aware of having.

Adam: What is it to be aware or conscious of an experience one is having? Whenever I see something am I conscious of seeing it? And when I hear something, am I conscious of hearing it? I may be conscious of *what* I see if it has caught and held my attention, but what is it to be conscious of my seeing whatever I see?

Christopher: Well, I'm not sure. Hell, I'm not a philosopher, I'm just a scientist. But even if you're right about what conscious experience is, I'm still not convinced that there isn't something it is like to be a conscious creature.

Adam: What aren't you convinced about?

Christopher: Well, it still seems to me that while there isn't anything it is like for a brick to be a brick, or for a laser-jet printer to be a laser-jet printer, there is something it is like for a human being to be a human being. And that's because human beings are conscious and have conscious experience. If a kind of creature has conscious experience then there is something it is like for that creature to be the creature it is. So, for example, there is something it is like to be a dog, to see the world from the dog's point of view – not to mention to smell the world from a dog's point of smell [*he chuckles*]. And presumably there is something it is like to be a bat, and to perceive the world from the bat's point of view, by means of echo-location. I mean, that's where I started out from this evening, and you haven't shaken my confidence in that!

Stranger: Well, we must clear this up, huh? Just as we examined the question 'What is it like for you to experience this or that?' and came up with some surprising results, so too we must examine the question 'What is to like for you to be a so-and-so?' And perhaps this too will surprise us, huh?

Bruce: Well, it seems clear enough that while we know what it's like to be human, we can never really know what it's like to be a dog or a bat. Because we can't have their doggy or batty [*he chuckles*] experiences.

Stranger: Bruce, I told you to open the gate, not to try jumping it. Hmm!

Bruce: [*laughs*] OK. You give me the key.

Stranger: *Ja*, let me try. Look, Bruce, what is it like to be a philosopher?

Bruce: Well, I dunno – it's not bad. It's a pretty interesting thing to do; the teaching is fun; the problems are good, even if the pay isn't.

Stranger: So you answer the question of what it's like to fulfil a certain role by describing the characteristic features of fulfilling the role, no? You give the pros and cons, *Ja*.

Bruce: Yeah, sure.

Stranger: Now, sometimes we need to add a further specification to the question. We ask, for example, 'What is it like for a woman to be a soldier?', for we want to know something about the experience *of a woman* in the army, as opposed to a man, huh? Or we might ask the new Wimbledon star 'What is it like *for a teenager* to be the champion?' Hmm.

Jocelyn: But we might also ask simply 'What is like *for you* to be something or other?', as when we ask our parent 'Mum, what was it like for you to be a soldier?' or 'Dad, what was it like for you to be a sailor?' And the answer may be 'It was great', 'It was awful' or 'It was very exciting', and so forth.

Stranger: Good, that is correct and important. Now let us examine the form of these questions, hmm.

Jocelyn: Well, I suppose that the general form is 'What is it like for an X to be a Y?', or 'What is it like for me, or for you, or for her or for him, to be a Y?'

Stranger: *Ja*, good. The general form of the question always involves a comparison class, that is, we ask what it is like for a woman, *as opposed to a man*, to be a soldier; or what it is like for a teenager, *as opposed to an older person*, to be a champion; or what it is like for a doctor, *as opposed to some other professional*, to be a prime minister; and so forth. Now think, what if we ask 'What is it like for a doctor to be a doctor, huh? Or 'What is it like for a soldier to be a soldier, huh?'

Jocelyn: [*hesitates*] Well ... It sounds odd. One can ask what it is like to be a soldier or a doctor or a philosopher. That is, just to ask for a description of the role, the hardships and satisfactions of these jobs, and perhaps the typical experiences and episodes in this kind of professional life. But it would be very odd to ask what it is like for a doctor to be a doctor.

Adam: Yes, that's right. But there's more awry than just oddity. To ask what it's like for a doctor to be a doctor doesn't make sense, because no one *other* than a doctor *can be* a doctor – as opposed to becoming one – and no one other than a soldier can be a soldier. And so on.

Stranger: Precisely, Adam. Hmm. And *so* too, we can ask what it was like *for you, as opposed to others*, to be a soldier, a sailor, a tinker or a tailor, *Ja*, haha. For then you would describe not the typical form of life of a soldier or sailor, tinker or tailor, uh. You would describe *your* specific experiences, joys and sorrows, strivings, successes and failures, huh? But, now, what about 'What is it like for you to be a human being, huh?'

Jocelyn: Ah, yes. I see. There is no principle of contrast.

Christopher: I don't follow.

Jocelyn: Well, could you be anything other than a human being?

Christopher: Well, I could turn into a frog or a giant beetle, like that character in Kafka.

Adam: Come now, Chris. You can imagine, fantasize, these fairy tales, but they don't make sense. If you were to change into a frog or a beetle, you would cease to exist. The frog wouldn't be you.

Christopher: I don't see why. What if the frog had all my memories?

Adam: Oh, come off it, old boy. What is that supposed to mean: 'What if the frog had your memories?'? What would it *be* for a frog to remember being a neuroscientist? What, so to speak, would it *do* with your memories if it had them? That's just tosh!

Stranger: Shush, now. We can explore that road some other time. Hmm. For the moment, let us accept this, hmm! *So*, Jocelyn, now carry on.

Jocelyn: Yes, … Well, a human being can't be anything other than a human being, just as a dog can't be anything other than a dog, and a bat can't be anything other than a bat.

Adam: And it's easy to see why, since the identity of the human being, the frog or the dog depends upon the continued existence of the substance they are – of the kind of animal they are. You might turn into a frog, but then *you* – our good friend Chris – would no longer exist.

Jocelyn: Yes, I *see*. So it does not actually make any sense to say that there is something it is like for me to be a human being, or for a human being to be a human being – I couldn't be anything other than a human being, and human beings could not be anything other than human beings. So the question 'What is it like for a human being to be a human being?' is either nonsense or it just means 'What is human life like?'

Stranger: Precisely. And we can answer that nebulous question in many rough and ready ways, no? – Crossed with hope and fear, huh? Or, as your Hobbes said, nasty, brutish and short, huh?

Adam: And the claim that there is something it is like for me to be me is just nonsense on stilts. For it means nothing at all to suppose that *I* might not have been me!!

Jocelyn: So the statement that there is something that it is like to be a bat only we can't say what it is like is nonsense too. I mean, we *can* describe the life of a bat, if we know enough about their lives and behaviour. There is nothing ineffable about that. So we can answer the question: what is the life of a bat like. But that doesn't mean that *there is something that it is like* to be a bat. Or a cat. Or a human being.

Bruce: Well, I'm not sure what to say. But where does that leave us when it comes to consciousness? I mean, the what-it's-likeness-of-experience was supposed to clarify the notion of consciousness. A conscious creature was supposed to be a creature that has conscious experience. And conscious experience was supposed to have a qualitative character – there was supposed to be something it is like to have it. Now you've pulled all that apart.

Christopher: Wait a minute. I still don't see. I mean, what I said still seems to contain some grain of truth. You remember I said that *there* *isn't anything that it's like to be a brick or a laser-jet printer, but there is something it is like to be a human being.* OK, you've shredded all the stuff about there being something it is like to be whatever, but it still captivates. Why?

Stranger: It is very simple now. You can ask a human being what it was like for him to do or to be the various things he has done or been. Hmm. And within the constraints we have examined, one can answer the question without any difficulty, huh? But you can't ask a brick what it is like for it to fill the gap in the wall it fills, and you can't ask the laser-jet printer what it is like for it to print your latest paper. It is not, *so* to say, as if the brick takes pride in filling a gap in the wall, or the laser-jet printer enjoys printing your papers, huh?

Christopher: You mean that all this thirty year's worth of excitement just boils down to the platitude that sentient creatures enjoy experiences, and inanimate and insentient beings don't?

Stranger: *Jawohl*, I am afraid that is correct. Although we may add that, of the experiences they enjoy, some are agreeable, others disagreeable, and most are neutral and colourless. Hmm. And we must not forget that although we may ask what it is like for an X to be a Y, or for an X to enjoy or undergo some specific experience, that does not mean that there is something it is *like*, but only – at best – that there is something that it *is* to do this or that, or to be this or that.

Bruce: But what about consciousness? I mean, is that all there is to it? Hell, y' haven't even mentioned self-consciousness. What is self-consciousness?

Stranger [*rising from his chair*] Well, Bruce, that is a subject for another evening. Adam, thank you very much for a very good dinner and the excellent cognac. And thank you all for an enjoyable conversation, hmm.

Adam: [*rising to his feet, and helping Jocelyn to hers*] Well, thank you, my friend; that really was very helpful. Bruce and Chris, will you close shop while I take Jocelyn and our friend to their rooms.

Jocelyn: Thank you so much. That was great fun. Goodnight!

Stranger and Adam: Goodnight!

[*The three of them exit. Chris and Bruce put the glasses on the sideboard and tidy the table*]

Bruce: I feel a bit battered.

Christopher: Yeah. Now I know what a floor cloth feels like.

Bruce: What d'you mean?

Christopher: Well, he wiped the floor with us.

Bruce: Well, he wiped the blackboard clean. And I can't see anything on it at all now. I'll have to digest all that.

[*They turn the lights out, and exit, closing the door behind them. A dim light shines on the bust of Socrates, slowly getting stronger. The sound of chuckling can be heard. As it fades away, the light fades too*]

Notes

1   The idea that experience is to be characterized in terms of what it is like to enjoy or undergo it was advanced by Thomas Nagel, in his paper 'What Is It Like to Be a Bat?' (repr. in his *Mortal Questions* (Cambridge University Press, Cambridge, 1979), pp. 165–80). He claimed that 'the fact that an organism has conscious experience *at all* means, basically, that there is something it is like to *be* that organism' (p. 166) – 'something it is like *for* the organism'. This he dubbed 'the subjective character of experience'. Every subjective phenomenon, he held 'is essentially connected with a single point of view' (p. 167), which is species specific. These ideas spread like wildfire, and were adopted by hundreds of philosophers, psychologists and cognitive neuroscientists.

2   A persuasive argumentative move made by Martin Davies in M. Davies and G. W. Humphreys (eds.) *Consciousness* (Blackwell, Oxford, 1993), p. 9.

3   John R. Searle, *Mind: A Brief Introduction* (Oxford University Press, Oxford, 2004), pp. 135–36.

4   This linguistic aberration is now widespread among the self-styled consciousness studies community. See, e.g., the numerous entries pertaining to consciousness in T. Bayne, A. Cleeremans and P. Wilken (eds), *The Oxford Companion to Consciousness* (Oxford University Press, Oxford, 2009), and papers in M. Velmans and S. Schneider, *The Blackwell Companion to Consciousness* (Blackwell, Oxford, 2009).

5   J. Levine, 'Explanatory Gap' in Bayne, Cleeremans, and Wilkins (eds), *The Oxford Companion to Consciousness*, pp. 279–80.

6   C. I. Lewis, *Mind and the World Order* (Charles Scribner's Sons, New York, 1929), p. 121.

7   Francis Crick, *The Astonishing Hypothesis* (Touchstone, London, 1995, p. 9).

8   Ibid., pp. 9–10.

9   The problem of the inverted spectrum was first raised by Locke in the *Essay*. The question is whether it is intelligible that what one person sees as red, say, another sees as green, and so forth, but the green he sees he calls 'red', and so forth. So the fact that the way he sees coloured things is, as it were, inverted is not detectable, since the vocabulary has been shifted correspondingly.

10   Searle, *Mind*, p. 134

11   D. Chalmers, *The Conscious Mind* (Oxford University Press, Oxford, 1996), p. 10.

12   C. Koch, *The Quest for Consciousness* (Roberts, Englewood, CO, 2004), p. 4.

13   See Dialogue Five: 'On the Objectivity or Subjectivity of Perceptual Qualities'.

14   See Wittgenstein, *Tractatus Logico-Philosophicus*, 3.263, where he ascribed to 'This [patch] is red' the role that properly belongs to 'This [colour] is red', a mistake he later corrected in his *Philosophical Remarks*, §6.

15 A view embraced by Wittgenstein in the *Tractatus* and later rejected.
16 Searle, *Mind*, p. 135.
17 See Dialogue Eight: 'Can You Have My Pain? Can Different People Have the Same Pain?'

SUPPLEMENTARY READING

M. R. Bennett and P. M. S. Hacker, *Philosophical Foundations of Neuroscience* (Blackwell, Oxford, 2003), chaps 9–12.

P. M. S. Hacker, *The Intellectual Powers: A Study of Human Nature* (Wiley-Blackwell, Oxford, 2013), chap. 1.

Norman Malcolm, 'Consciousness and Causality' in D. M. Armstrong and Norman Malcolm, *Consciousness and Causality* (Blackwell, Oxford, 1984), pp. 1–101.

A. R. White, *Attention* (Blackwell, Oxford, 1964), chap. 4.

SECTION 3

A DIALOGUE ON THE OBJECTIVITY OR SUBJECTIVITY OF PERCEPTUAL QUALITIES

INTRODUCTION

The thought that colours, sounds, smells, tastes, heat and cold, wetness and dryness, and so forth are essentially subjective and relative was advanced by Galileo in the seventeenth century. These so-called secondary qualities were, he suggested, inessential to the notion of a material body. They play no role in a properly scientific account of the material world, and, by contrast with primary qualities such as shape, solidity, size and weight they are causally impotent. In fact, they are more akin to pains and tickles than to shape or motion. Nature in itself is colourless, soundless, tasteless and odourless, neither hot nor cold, neither wet nor dry. Secondary qualities are produced in our minds as a result of the impact of bodies or waves upon our sensitive organs, just as pains are produced by cutting our flesh. They are essentially subjective. Remove sentient creatures from the world, and all is dark. Thunderstorms produce no noise – unless there are ears to hear, and trees on desert islands fall soundlessly to the ground – unless there is a Robinson Crusoe to listen.

This conception won the allegiance of the advanced scientists of the day including Boyle and Newton. In our times, it is the received view of most psychologists, cognitive neuroscientists and physicists. Many consider this conception to be a deliverance of science, resting firmly on empirical evidence. But to be sure, there is no experiment to prove that grass is not green or that roses are colourless. The claim that the so-called external world is, in itself, colourless is a piece of seventeenth-century metaphysics, not of twenty-first-century science. Although this conception is now conventional wisdom, and indeed is taught to schoolchildren as such, we have lost the sense of shock and wonder at it. For, of course, if it is true, we live in a world of illusion – nothing is really, in and of itself, coloured; there aren't really any sounds at all – the music is all in our minds or brains. Lilac and jasmine do not really have a scent – unless they are smelled. Fire may, to be sure, have a temperature, but it is not really hot, and ice is frozen water, but it is not really cold. And so on. The world as we perceive it is just the stuff that dreams are made of. Our minds, so to speak, should reel at the very thought.

Seventeenth-century philosophers moved along parallel tracks. Descartes embraced this Galileian conception of reality with alacrity. The material world, he thought, is characterized by matter, the essential characteristics of which are mathematical. Having redrawn the boundaries between the mental and the physical, he advanced a highly Pythagorean conception of the material world. Reality is subject to quantification and measurement. A mark of objectivity in the material world, he held, is being amenable to mathematical description and being subject to physical laws that can be mathematically described. This conception was adopted and further elaborated by Locke, whose discussion of primary and secondary qualities became canonical.

It is remarkable how easily we get drawn into this strange seventeenth-century metaphysical conception of the nature of the world around us, and of the subjective character of the world as we experience it. It is no less extraordinary how difficult it is to shake oneself free of this picture of reality and of our ability to apprehend it as it is. The following dialogue in Elysium invites the reader (or listener) to witness an impassioned debate on this theme, in the hope that it will stimulate further thought and reflection. Dr James Lockett is an eighteenth-century follower of John Locke (the reason Locke does not appear himself is that I did not want to get embroiled in academic disputes over the intricate details of his conception). Dr Lockett advances the standard eighteenth-century philosophical conception of the matter: secondary qualities as we perceive them are no more than ideas in the mind caused by 'external objects'. Professor Ronnie Freiberg, an American neuroscientist, presents the viewpoint of the majority of contemporary cognitive neuroscientists, psychologists and physicists: secondary qualities as we perceive them are creations of the brain, not qualities of things in the world. Professor Timothy Swan is an Oxford don, who advances views that have been current among Anglophone philosophers in general, and Oxford philosophers in particular, for the last three or four decades. Thomas Roe is an intelligent Oxford undergraduate, who is rightly amazed at what his elders tell him and tries to challenge them. In this endeavour, he is greatly assisted by Socrates, who, as ever, asks questions, challenges received views and rejects shoddy arguments. Being in Elysium, he has, as will become evident, enjoyed many long and interesting discussions with his friend Wittgenstein, from whom he has learnt much.

Fifth Dialogue

ON THE OBJECTIVITY OR SUBJECTIVITY OF PERCEPTUAL QUALITIES

'The picture is something like this: Though the ether is filled with vibrations, the world is dark. But one day man opens his seeing eye, and there is light' (Wittgenstein)

Protagonists:

Thomas Roe: a bright twenty-year-old Oxford undergraduate; dressed as a (respectable) modern-day student.

Professor Ronnie Freiberg: an American scientist presenting a Galtonian picture of views advanced by physicists, psychologists and neuroscientists; in his early forties; dressed in an open-necked shirt, jeans and comfortable loafers; very laid back, friendly and informal.

Dr James Lockett: an elderly gentleman, somewhat shy and diffident; pupil of Locke; meticulous and pedantic pronunciation; wearing an early eighteenth-century dress of a scholar, bewigged.

Professor Timothy Swan: Tom's tutor, a middle-aged Oxford don, presenting a synopsis of views current among Oxford philosophers over the last four decades; dressed in a tweed jacket and tie, casual light brown slacks; neutral educated English accent; slightly pompous and condescending.

Socrates: gruff gravelly voice, slight regional accent; wearing Greek dress.

The scene is a garden in Elysium. The sun is shining. A rich verdant lawn is surrounded by flower beds and rose bushes in bloom, with tall trees behind. Beyond, there is a beautiful view of lake and mountains. Five comfortable garden chairs are placed in the shade beneath some trees. There is a low table on which there is a wine decanter and glasses of wine, and some scattered books.

Dr Lockett, Professor Swan and Professor Freiberg are arguing with Tom. One chair is vacant.

Tom: But you can't really mean that! You can't really mean that all *this* [*he gestures at the garden all around*] is just a kind of, kind of, *illusion* – that nothing here is coloured!

Professor Freiberg: Yeah, that's exactly what we mean. Color's just a sensation in the brain. Science has proved this, Tom. What colors things seem to have is a function of the spectrum of light they reflect. Things absorb or reflect electromagnetic radiation that impacts their surface. If the radiation is in the 390 to 700 nanometer wavelength, it's what we call 'visible light'. If a surface absorbs all the light that impacts it, it appears black. If a surface reflects and scatters it all, it looks white. All the chromatic colors are produced by partially reflected light of one or another band of wavelengths. So things that look red reflect light of 700 to 635 nanometers, and absorb the rest. We experience a sensation of red because the retinas of our eyes contain three types of color receptor cells, that is, three types of cone: S, M and L cones, responsive to short, middle and long wavelengths. The tristimulus values are conveyed from the eyes to the brain in the form of neural impulses by three different channels: a red/green channel, and blue/yellow channel and a black/white luminance channel. The upshot of all this is that we experience color sensations in the brain. So colors are purely subjective sensations. OK, I know it sounds odd. But it's true.

Dr Lockett: And not Colour alone, my young Friend. Taste, Smell, Sound, Warmth and Cold, all these *secondary* Qualities, as we denominate them, are not Real. They are nothing but a certain sort and degree of Motion in the minute Particles of Matter.

Tom: You mean that the sun is not, like, really shining, that those roses have no scent, that this wine [*he takes a sip from the glass on the table before him*] is tasteless, and that it is not warm today? Isn't that ... sort of ... absurd?

Dr Lockett: No, no. Messer Galilei showed that natural Philosophy has no use for such Qualities as we deem Colours, Sounds, Tastes, Smells, Warmth and Cold to be. If the perceiving Creature were removed, he said, all of these Qualities would be annihilated and abolished from existence.[1]

Tom: Dr Lockett, I don't understand. What d'you mean 'natural philosophy has no use for such qualities'? Even if it has no use for them, why should that imply that they don't, kind of, really exist?

Dr Lockett: Galilei showed that *every Particle of Matter must be Solid, have Extension, Figure and be in Motion or at Rest* – these real Qualities being what we called 'original' or 'primary' Qualities. For take anything, perchance a Grain of Wheat. Let it be divided until its smallest Parts are obtained. These Parts still retain their Solidity, are extended, possess some Figure and Bulk, and are either at Motion or at Rest. But as for Taste, Smell, Colour or Sound, these, as Master Locke taught, in truth are nothing *in the Objects themselves*. For the Corpuscles in themselves lack Colour; they make no Sound; they have no Taste and they are not fragrant. Secondary Qualities are nothing but *the effect on us* of the Structure and Arrangement of the Particles – the Corpuscles – of material Things. As we perceive them, they are but Sensations in the Mind.

Tom: I don't see why.

Professor Freiberg: Look Tom: Galileo showed that you can formulate the laws of physics without mentioning secondary qualities. Newton's great theory confirmed this. And although we have replaced Newton's theory by relativity theory and quantum mechanics, we've seen no reason to change our view on this. Colors, sounds, smells and tastes have no explanatory role in physics, *because they are causally impotent*. Have you gotten hold of that? [*Tom nods*] OK.

Now, as Dr Lockett was explaining, secondary qualities are not *dissective* – that is, when you divide any lump of stuff, you'll get to a point where the particles no longer look colored, sound noisy, taste sweet or sour, feel hot or cold. So atoms, for example, aren't colored or noisy, hot or cold, and so on. OK so far? [*Tom nods*]

Right. So given that atoms and molecules have no color, don't make any noise, and don't have a smell and so on, Galileo, Descartes, Boyle and Locke all thought that we can explain *all our perceptions* without even mentioning secondary qualities in the explanation. Now as it turned out, their intuition was right. If you wanna know why we have an experience of redness when we look at the rose here, it is because the surface of its petals reflects light in the 635–700 nanometer band, and absorbs the rest of the light. Right! Now, what we've found is that in explaining our perceptions of color, all we need to mention are the reflectance properties of surfaces and the radiation that impacts our retinas. The rest is neuroscience.

Tom: What do you and Dr Lockett [*he turns towards Lockett*] *mean* by 'a red sensation' or 'sensation of red'?

Dr Lockett: Let us consider Sensations caused in us by Bodies in the external World. Water, we say, is hot or cold. That is to say, it has, by the Motion of its minute Particles, the Power to cause in us a Sensation of Warmth or a Sensation of Cold. Now is it not evident that there is nothing in the Bodies themselves but the agitated Motion of Particles – which causes a Sensation of Heat in us? There is here a similitude with Pain. For consider a Knife that cuts our Flesh. We feel an immediate Pain. But there is nothing in the Knife that resembles the Idea of Pain in us. From whence I think it is easy to draw this Observation, that so it is likewise with Colours and Sounds. Sir Isaac Newton, in his never enough to be admired Book *Optics*, demonstrated that Colours are but Sensations in the Sensorium. Look, we have his Book here [*he leans forward a picks up a copy of Newton's* Optics *from the table, and opens it*], and I have marked the Place wherein he states his conclusions. Here it is:

> a Sound in a Bell or musical String or other sounding Object, is nothing but a trembling Motion, and in the Air nothing but that Motion propagated from the Object, and in the Sensorium 'tis a Sense of that Motion under the Form of Sound; so Colours in the Object are nothing but a Disposition to reflect this or that sort of Rays more copiously than the rest; in the Rays they are nothing but the Dispositions to propagate this or that Motion in the Sensorium, and in the Sensorium they are Sensations of those Motions under the Form of Colours.[2]

[*He closes the book and puts it back on the table with a smile*] Now, is anyone able to gainsay this?

Professor Freiberg: Yeah, that's great stuff. Of course, we know a great deal more about light, light reflectance value, the way the eye functions, the optic nerves and the chiasma and about the visual striate cortex than Newton did. But he sure was on the right route.

Professor Swan: *And* we have greater metaphysical knowledge than they had in the seventeenth century – or in the twentieth for that matter. We have at last reached the end of the beginning in philosophy, and are making greater progress than ever before [*he smiles with satisfaction*].

Tom: So you really *do* mean that there aren't any colours in the actual world? That there are no sounds, or smells or tastes? Then … nothing is what it seems to be. We are living, kind of, in perpetual illusion! Nothing is really coloured in the world around us, and things don't really make sounds. Nothing is really warm or cold. And all food, in itself, is, like, tasteless! I just can't believe that. I mean, doesn't this scientific story strike any of you as … as kind of *weird*?

Professor Freiberg: Well, that's what modern scientists have shown. Sir John Eccles, a fine neuroscientist, who won the Nobel Prize, said that there's no color in the material world, only the emission of electromagnetic waves. My neuroscientist colleague in California, Eric Kandel, who also got the Nobel, wrote a great passage too. Here [*he leans forward and picks up another book and flicks through it until he finds the passage he is looking for*] Yeah, here it is.

> We *receive* electromagnetic waves of different frequencies but we *perceive* colours: red, green, orange, blue or yellow. We receive pressure waves but we hear words and music. We come in contact with myriad of chemical compounds dissolved in air or water but we experience smells and tastes.
>
> Colors, sounds, smells and tastes are mental constructions created in the brain. Therefore, we can ask the traditional question raised by philosophers: Does a falling tree in the forest make a sound if no one is near enough to hear it? We can say with certainty that while the fall creates pressure waves in the air, it does not create a sound. Sound occurs only when the pressure waves from the falling tree reach and are perceived by a living being.
>
> Thus, our perceptions are not direct records of the world around us but are constructed internally according to innate rules and constraints imposed by the capabilities of our nervous system.[3]

And I recollect Semir Zeki, whose lectures I was privileged to audit in London, saying that color is a property of the brain, not of the world outside.[4]

Professor Swan: [*slightly patronizing*] Look, Tom, this is the received scientific view of the matter. Of course, it needs some philosophical refinement. But you don't really want to defend *common sense* in the face of science, do you? After all, science is our best way of achieving knowledge about the world; it gives us the most reliable picture of how things really are. Surely the weight of all this authority is enough for you.

Tom: Well, I'm not sure … I *was* kind of defending common sense, and common sense at least makes sense. What's wrong with defending common sense?

Professor Swan: Well, philosophy should not pick quarrels with science. After all, philosophy is just an extension of scientific investigation. It is at one with science in its quest for knowledge of the world, even though it pursues it at a higher level of generality. But there is no need to offend common sense by saying that fire isn't really hot, nor snow really cold. Actually, Lockett, your teacher Locke was more careful, was he not?

Dr Lockett: Master Locke was indeed a cautious Philosopher, not given to rash or hasty pronouncement. He averred that secondary Qualities *as we perceive them* are but Ideas in the Mind. But he was averse to claiming, against Reason, that Snow is not white nor Fire hot. Rather, said he, Flame is denominated 'hot' and 'light', and Snow 'white' and 'cold' *from the Ideas they produce in us*. To discover the Nature of these Qualities better, and to discourse of them intelligibly, it is convenient to distinguish them as they are Ideas or Perceptions in our Minds and as they are Modifications of Matter in the Bodies that cause such Ideas or Sensations in us. For we may not think, with the Vulgar, that the Ideas within our Mind, our Sensations of Heat and Cold, of Pain and Pleasure, of Colour and Sound, are exactly the Images and Resemblances of something inherent in the Objects themselves. Most of these Sensations in the Mind are no more likenesses of something existing without us than the Names that stand for them are likenesses of our Ideas. Rather, Master Locke explained, secondary Qualities in truth are nothing *in the Objects themselves* but *Powers* – Powers to produce various Sensations in us by the primary Qualities of their insensible Parts.

Tom: You mean that names of secondary qualities are kind of ambiguous? Surely that is not what *you're* saying, Professor Swan?

Professor Swan: Well, ... roughly. Or rather, yes, but in a rather more ... er ... refined and precise way ... semantically speaking, of course.

Tom: So the word 'red' sometimes means a sensation, sometimes a reflectance property of objects and sometimes the structural arrangements of molecules on the visible surfaces of things? The one thing it doesn't seem to mean is: the visible colour of those ☞ roses, or of blood, or of ripe tomatoes!

Well! ... I'm really confused now. With all these arguments from scientists and philosophers, I suppose it *must be like this*. But when I look around, hear your voices, smell the roses and taste the wine, I know it, kind of, *can't be like this* ... I give up. I don't know what to say.

[*While Swan was talking, Socrates, strolling by in the background, stopped to listen. Now he comes forward*]

Socrates: No, no, young man. This is just when you should not give up. [*He approaches them*] Good afternoon to you all. May I join you?

[*Professor Freiberg jumps up and offers Socrates the vacant chair*]

Professor Freiberg: We'd be honored, Socrates.

[*Socrates sits down in the vacant garden chair*]

Socrates: Thank you … What a strange coincidence. I was talking to my friend Ludwig yesterday – a very interesting man, by the way. Full of ideas. Full of questions. Wonderful. He said something very similar to what you just said, Thomas.

Tom: What did I just say?

Socrates: You said that things must be like this, and yet on the other hand, they can't be like this. Wittgenstein was telling me what captured his imagination about philosophy. He said that it was the general form of deep philosophical problems.

Tom: What's the general form of philosophical problems?

Socrates: Just what you said: when we look at the problem it seems to us, on the one hand, that things *must* be so, and, on the other hand, that things *can't* be so. But this is not the point at which we should give up. It is at this point that we know that we should persevere. For we know we are in a muddle, and that we have tied a knot in our understanding. So we must untie it. Now, Thomas, tell me what is the matter.

[*A pause while Professor Freiberg pours Socrates a glass of wine. Socrates drinks the whole glass with relish, and smiles with pleasure*]

Socrates: Thank you. Excellent.

[*He turns to Tom, waiting for him to speak. Tom takes a deep breath*]

Tom: Well, it's like this. Ronnie, Dr Lockett and Professor Swan were all trying to persuade me that colours, sounds, smells, taste, warmth and cold – what they called 'secondary qualities' – are really just, kind of, subjective appearances. They say that if you were to take away all sentient creatures, then there would be no colour in the world. Actually, they even say that when a tree falls on a desert island it doesn't make a noise, it just makes sound waves in the air. And roses have no smell either. They say that science proves this.

   Now, according to Ronnie, there aren't any objective colours or sounds, tastes or smells. Then Dr Lockett and my tutor, Professor Swan, said that this wasn't quite right. They said that 'red', for example, sometimes means a subjective sensation, but sometimes it means something objective. But I didn't quite understand what they think it means objectively. Dr Lockett said that it was a structure or arrangement of atoms on the surface of things; but then he went on to say that it was a power to cause sensations of red in us. Professor Swan said that red as it is in objects – that's the phrase they both used: 'as it is in objects' – is just a disposition in things. I think what he means by 'disposition' is what Dr Lockett means by 'power'. So according to them, 'red' seems to mean three different things – although I must say, *I've* never noticed three different meanings. I think 'red' means a colour that you see – like the roses on that bush. I mean ... the roses are red, aren't they? You can see that they are. But they all tell me that I am being naïve and unscientific. And then they said that physics has no need of secondary qualities to describe and explain events in the world, so they are just subjective and relative to us. I couldn't follow that bit.

Socrates: Thank you, Thomas. I am afraid, gentlemen, that I know very little about these matters, and you must instruct me about your discoveries.

Professor Freiberg: It'd be an honor to do so, Socrates.

Professor Swan: [*to Professor Freiberg, sotto voce*] You don't know what you're letting yourself in for.

Socrates: But now tell me, is this really the last word in natural philosophy? Is this what has been discovered in the so-called twentieth century?

Professor Freiberg: Yeah, sure!

Socrates: That is strange. It all sounds to me to be very like the ideas I heard from that Abderite philosopher ... what was his name? ... Ah, yes – Democritus. I talked to him when he visited us in Athens. Of course, no one knew him then. But he was a cheerful fellow – laughed a lot, you know. I enjoyed our talks in the agora. He said that things are sweet or bitter by convention, hot or cold by convention and coloured by convention too. But in reality, he insisted, there is nothing but atoms and the void. I couldn't understand what he meant by the phrase 'by convention'. Of course, when I asked him, it emerged that by 'by convention' he really meant 'relative to us' – funny way of expressing oneself!! ... Still, his idea that colours are relative to us, or that colours are not objective or that colours are not real is worth pursuing ... Hmm! So things don't really have a colour at all! And don't make any sound either. Fire isn't really hot, and ice isn't really cold! ...

Hmm! All very confusing, this atomism ... Still, as I was saying, nothing much *new* about this view.

Professor Freiberg: Well, may be this guy Democritus thought up an atomistic hypothesis in your days, Socrates. But it wasn't until the nineteenth century that we could prove it all. I mean, Dr Lockett, when your teacher John Locke was advancing his ideas, and when Newton wrote the *Optics,* there wasn't any experimental proof that matter consists of atoms bonded together to form molecules, was there?

Dr Lockett: No, no, indeed not. No, no. It was a most excellent Hypothesis, which seemed to explain much. But in truth, we had no *Experimentum crucis* to prove it. So it was with much Joy that I came to learn that our Successors had, with the passing Years, proved it to be true.

Socrates: Very good. However, we were not speaking of atomism, but of what you call 'secondary qualities', such as colours. Did you prove that colours are, as you say, but creatures of the mind?

Professor Freiberg: Yeah, sure; well, of the brain at any rate.

Socrates: I apologize for my ignorance. Pray tell me, what is the proof?

Professor Freiberg: Well, perhaps there isn't exactly a proof, but everything speaks for it, y'know, and ...

Socrates: Look, my friend, is there, or is there not, a decisive experiment that you can perform before us that will show that things as they are independently of us, are not coloured, or noisy, or warm or cold? Or that these ☞ roses have no scent [*he reaches for a rose and smells it*]? Ah, exquisite. So, is there an experiment that shows that nothing around us is coloured or fragrant?

Professor Freiberg: Well, ... no, not exactly, not an experiment.

Socrates: So even if you have a multitude of explanations of how we perceive things, and descriptions of the invisible structure of the surfaces of things, you can't actually *prove* any more about the *unreality* of what you call 'secondary qualities' than my old acquaintance Democritus could.

Professor Swan: Well, I don't think ...

Socrates: Wait a moment, Professor Swan. I have another question to put to you all. You say that those roses aren't *really* red – at least, not in the sense in which we think they are, that the grass is not *really* green and so on. But to lack a colour is surely to be colourless is it not?

Tom: That's right.

Socrates: And the water in that ☞ jug is colourless.

Professor Swan Yes, of course, but ...

Socrates: And these glasses too [*he holds up his empty glass*] are colourless, are they not?

Professor Freiberg: Oh, I *am* sorry [*he fills Socrates's wineglass from the decanter*]

Socrates: Thank you. [*Drinks*] Aah! ... Now, when you say that nothing is coloured, you don't mean to say that everything around you is colourless and transparent, like glass and water.

Professor Swan: No, of course not. That would obviously be false.

Socrates: Quite so. Now, will you grant that whatever is colourless *could be* coloured? The water in the jug will become red if we add wine to it. Glass will become blue if the glassblower adds copper to the molten glass.

Tom: Yes, yes. That's cool.

Socrates: So you don't want to say that everything is colourless in the ordinary sense of the word. And you don't want to say that anything is coloured in the ordinary sense, either. So what do you want to say?

[*Short silence*]

Professor Swan: Well, what we are saying is that it is a *metaphysical* truth that colours are subjective modifications of the mind, or to be more accurate ...

Socrates: Slow down! Slow down! You are going much too quickly for me. [*He chuckles*] As a friend of mine said the other day, 'Philosophers should greet each other with the words "Take it slowly!"'.

An unfortunate term 'meta-physics' – it's all the fault of Aristotle's bungling editor Andronicus of Rhodes. I've never understood what exactly it means. The study of Being as Being? ... Humph! Parmenidean twaddle. Never mind!

At any rate, it seems clear that what you are offering young Thomas here is not the latest news from the laboratories of natural philosophy, but tired old news from philosophy. Haha! From meta-physicists, haha!

Professor Swan: Socrates, that is hardly ...

Socrates: No, no. Let's be serious. Have I understood you correctly: do you mean that colours are essentially properties of our ideas and impressions?

Professor Swan: Well, you could put it like that, although it is rather old fashioned. In a more modern and more precise idiom, if you prefer, they are representational properties of the representational contents of visual experiences.[5]

Socrates: I prefer not, hahaha, thank you. Maybe we'll come back to that. In the meantime we have quite enough trouble. So you don't just think that objects *aren't* coloured, you think that objects *can't* be coloured, or noisy, or smelly – as we take them to be, I mean?

Professor Swan: Yes, that's right. They can be coloured and so forth only in the sense that secondary qualities, as they are in objects, are just dispositions to cause representations in us.

Socrates: Yes, yes, we'll come to that bit in a moment. So you think that before living creatures opened their seeing eyes, the ether was full of vibrations, but all was dark.[6] Then the gods created life, and the sun shone.

[*Short silence*]

Dr Lockett: [*taken aback*] Oh. No. Or perhaps, yes. How curious. I am uncertain. I had never thought of that.

Socrates: You must also think that material objects are like numbers?

Dr Lockett: [*flustered*] Must I? No, no, I do not. I mean, what *do* you mean?

Socrates: Well, do you think that the number three can be green, or the number four red?

Dr Lockett: No, Men, versed in philosophical Enquiry, should not speak so. It is evident that it is Numerals, not Numbers, that can be coloured – indeed, are coloured. But, note well, Numbers are not said to be colourless either. For they are not alike to Glass.

Socrates: Quite so. So you don't just hold that physical things *are* not coloured. Rather, it seems, they *cannot* be coloured. Tell me, are you then claiming that it makes no more sense to say that physical objects are coloured than it makes sense to say of numbers that they are coloured?

Dr Lockett: Oh. I hesitate ... Yes. I must allow it.

Socrates Now, tell me, why can't numbers be coloured?

Dr Lockett: That is evident. It is because they are not extended, and not visible either. One cannot see Numbers.

Socrates: Perhaps. But now, what of physical objects? Are they not extended?

Dr Lockett: Yes, I allow that.

Socrates: And are they not visible?

Dr Lockett: Well, yes. I must allow that too.

Socrates: What do you mean, 'I must allow it'? Can't you see the trees around us, the sward beneath our feet, the roses?

Dr Lockett: I concede that I can.

Socrates: But if they were neither coloured nor colourless and transparent, you would not be able to see any of them, would you?

Dr Lockett: Why no. I concede that too.

Socrates: I should hope so, too – lest you should have taken leave of your senses. Colours are visibilia, or they are nothing.

Dr Lockett: Yes, I must grant you that.

Professor Freiberg: No, no. You shouldn't have conceded that, Dr Lockett. Colors are sensations in the brain.

Socrates: I see; ... sensations in the brain. That is very interesting. You must tell me more, Professor Freiberg. I know that itches and tickles are sensations, which may be in the small of the back – which I always find difficult to scratch. Haha! I know that pains, such as headaches, are sensations. Are you saying that seeing colours is like feeling an itch, or that it is like having a headache?

Professor Freiberg: No, of course not. Colors are *visual* sensations. A headache is a *pain* sensation.

Socrates: So where do you see visual sensations?

Professor Freiberg: In the visual cortex, in the occipital lobes.

Socrates: With your eyes? You mean you peer into your brain to see colours?

Professor Freiberg: No, of course not.

Socrates: So you see colours without eyes?

Professor Freiberg: Well, no. You can't see without eyes.

Socrates: Quite so. Now, [*turning to Professor Swan*] Professor Swan, tell me something else. Is pain a sensation?

Professor Swan: That is what is being argued.

Socrates: And you feel pains.

Professor Swan: Yes, of course.

Socrates: Where do you feel them?

Professor Swan: Well ... anywhere in my body.

Socrates: And when you have a pain in your knee, what do you do?

Professor Swan: I suppose I rub it or assuage it, put hot compresses on it.

Socrates: So, if colours are sensations, they would have to be located somewhere in your body.

Professor Swan: Yes, they are in the brain.

Socrates: Are brains then red, yellow, green and blue?

Professor Swan: No, no, no. Colour is a property of visual sensations, or more accurately ...

Socrates: How could a sensation be coloured? I know how to colour my stone carvings, but how would I go about colouring my sensations?

Professor Swan: Oh, no. You can't *colour* sensations. Rather, just as you have somatic sensations *of* pain, so too you have visual sensations *of* colour.

Socrates: And when you have such a visual sensation, you have it in the brain?

Professor Swan: Yes, of course.

Socrates: Well, I am ignorant of such matters. But I seem to recollect someone telling me, perhaps it was the great neurosurgeon Mr Penfield, that there is no feeling in the brain. There can't be any sensations in the brain, because there are, as Mr Penfield explained to me, no fibre endings there except in the dura. Is that not correct, Professor Freiberg?

Professor Freiberg: Yeah, you're right. [*Sotto voce*] I can see where this is heading

Socrates: Of course, one can have a headache. So I suppose that colour sensations must be a sort of headache.

Professor Swan: [*with an exasperated sigh*] No, of course not.

[*Short silence*]

Socrates: No ... Still, I agree that there are visual sensations.

Dr Lockett: What do you mean, Socrates? I thought that you had just demonstrated that there are none.

Socrates: Not at all. I merely questioned whether seeing the colour of a thing is having a visual sensation. Indeed, there are auditory, olfactory and gustatory sensations too.

Dr Lockett: Aha! So you withdraw. You concede defeat!

Socrates: But, my friend, I have nothing to withdraw *from*. And I have nothing to concede, only to ask. No, no. But I should like you to tell me whether, when you come out of a dark room into the brilliant noonday sun, you are not dazzled?

Tom: Yes, of course, Socrates.

Socrates: And what do you do?

Tom: I rub my eyes, I s'pose. I shield my eyes. That sort of thing.

Socrates: So is that not a visual sensation – a sensation of glare? of being dazzled?

Tom: Yes, I guess so. I mean, I rub my eyes, just as I rub my knee if I bang it.

Socrates: So this visual sensation, the sensation of being dazzled, is it in your brain? Or in your mind?

Tom: No, of course not. The sensation of being dazzled is in my eyes – that's why I rub them.

Socrates: Is this visual sensation then a sensation of colour?

Tom: No, no. Certainly not. Obviously not.

Socrates: Are such visual sensations as being dazzled involved in seeing things, as pain sensations are involved in injury?

Tom: No. Strong sunlight, after one has been in the dark, blinds one for some moments. I mean, when I am really badly dazzled, I can, like, hardly see anything.

Socrates: Quite so. And when someone blows a trumpet close to your ear, you are deafened. *That* might be called an auditory sensation, and it hinders hearing sounds no less than being dazzled hinders seeing sights. And if you eat heavily spiced food, as the Persians are said to do, does that not numb your mouth and benumb your sensibility?

Tom: Yes, that's right.

Socrates: May we then not safely conclude that colours are not sensations in the brain? And neither are sounds and smells? Nor, of course, are they perceptions, for you must grant that *what* we perceive is not in the brain, and *our perceiving* what we perceive is not something that occurs in the brain, but – at the moment – in the garden. For if I ask you this evening where you saw and smelled those fragrant roses, will you not reply 'In the garden', rather than 'In my brain'?

Tom: Wow! That's cool!

[*Silence*]

Professor Swan: Yes. That is why I tried to put things more precisely, but you stopped me, Socrates.

Socrates: Forgive me, Professor Swan. We should certainly welcome more precision in these matters. Pray enlighten us.

Professor Swan: Gladly. [*He puts his finger tips together before continuing*] Perceptual experiences represent the environment of the perceiver in a certain way. A *visual* perceptual experience enjoyed by someone walking in the garden may represent the lawn, the trees and the flowers as having spatial relations to one another and to the experiencer, and as having various qualities. [*He gradually slips into lecture mode*] The representational content of a perceptual experience is described by a proposition or set of propositions, which specifies, or specify, the way the experience represents the world to be. Representational properties are properties an experience has in virtue of features of its representational content. [*His listeners' eyes are starting to glaze over*] In the representational content of an experience, objects are presented under perceptual modes of presentation. Now, it is in the nature of representational content that it cannot be built up from concepts unless the subject of experience himself has those concepts: the representational content is the way the experience presents the world as being, and it can hardly present the world as being that way if the subject is incapable of appreciating what that way is. So we must further distinguish properties from modes of presentation of properties.[7] [*Silence*] ... Now is that not clear? ... [*short silence*]

Socrates: Well ... perhaps not entirely. Of course, I can see what you mean by precision. But there were a few fine points that escaped me. First I should like you to explain to me what a representation is.

Professor Swan: Yes, of course. A painting of an actual landscape is a representation of the landscape it depicts. A portrait drawing is a representation of its sitter – a likeness. It represents what it is a portrait of. And of course, descriptive sentences in speech are representations – linguistic representations – of the states of affairs they describe.

Socrates: Now tell me, does a representation not require a medium of representation?

Professor Swan: Yes, of course. A painting requires oil paint, gouache or water colour; the sketch requires pencil, charcoal or silver point; and speech requires a voice of some pitch and tone; and so forth.

Socrates: And must not the medium of representation possess perceptible features that are not representative.

Professor Swan: What do you mean?

Socrates: Well, a representational painting consists of paint, although the paint itself does not represent anything. The lines drawn in a sketch must have some colour, although their colour does not represent anything. And the speed with which one talks does not represent anything, any more than the loudness of one's voice. You must surely grant that?

Professor Swan: [*warily*] Yes.

Socrates: And do you not admit that but for the non-representative qualities of the medium of representation, we would not be able to perceive the representation at all? For one cannot have a painting with neither paint nor support, or a pencil sketch on paper with no pencil marks or paper.

Professor Swan: [*reluctantly*] Yes, I suppose you are right.

Socrates: But now you must admit that your perceptual experiences do not have any *representational* content.

Professor Swan: I don't see why.

Socrates: Because your perceptual experiences, as opposed to their description, have no medium of representation. So they have no *non-representational* qualities in virtue of which they *can* be representations. Indeed, they are not perceptible at all. For you cannot perceive your perceptions – you have them. But it is the nature of a representation to be perceptible. Perceptions are not representations of any kind, and they can have nothing that could rightly be called 'representational content'. For only a representation could have a representational content.

Tom: Wow!

Socrates: And if something is a representation of a thing, my friend, then must one not be able to compare the representation with what it represents in order to see whether it represents accurately or inaccurately?

Professor Swan: Yes, of course.

Socrates: But how would you think to compare this internal representation of yours with what it allegedly represents? Can you hold up your perceptual experience, look at it and at what it is an experience of, in order to see whether it is accurate or not? Is that how you determine whether you have perceived what you perceive correctly or not? Is your perception of this garden like a picture that represents the garden and can be compared with it?

[A long silence, while Socrates beams around, pours himself another drink and quaffs it]

Professor Freiberg: No. No. We left the tracks way back along the line! Look! Seeing colors is a visual experience. Hearing sounds is an auditory experience. Colors, sounds, smells, tastes, hot and cold are not in objects at all. They are subjective features of consciousness. That doesn't mean that things around us aren't colored. But they are not colored *in the same sense*. For something to be red is for it to have a disposition to cause an experience of red in us. For something to make a noise is for it to produce sound waves that cause us to have an auditory experience. Isn't that right, Professor Swan?

Professor Swan: Yes, yes.

Socrates: So colour words are systematically ambiguous.

Professor Swan: *[by now a little upset and faintly aggressive]* Why not? Did not Frege say just that? – that they have a subjective sense in which they refer to sensations, and an objective sense in which they refer to what causes us to have that sensation.[8]

Socrates: But ambiguity is a local linguistic feature, is it not?

Tom: I don't understand that, Socrates.

Socrates: What I mean is that word-ambiguity is an accident that occurs in a language, when one and the same word is used to mean two quite different things – but it is not to be expected that ambiguity will reappear on translation. 'Port' is ambiguous in English, but not in other languages.

Professor Freiberg: You've lost me, Socrates.

Socrates: Well, according to you, 'red' has at least two different meanings. It means either a disposition to reflect light of certain frequency onto our retinas, or the quality of a subjective experience. 'Salty' means either a disposition to affect our tastebud receptors, or the quality of a gustatory experience. And so on. Now, does it not strike you as strange that 'κόκκινο' and 'red', and doubtless corresponding words in other languages too, must, on your view, share the same accidental ambiguity?

Professor Swan: No, not really. *This* ambiguity is not accidental. It is the product of precisely the error theory that we are advocating. As my colleague John Mackie used to say in his lectures, ascribing colour, as we see colour, *to objects* is all a mistake, a systematic error.[9] This is a systematic mistake made by the whole of mankind. Small wonder that they should employ ambiguous colour words in all languages.

Socrates: Is it really so? Tom, do you mean by the word 'red' a sensation in the mind or brain.

Tom: No, of course not. That was what I was kind of saying to them all along!

Socrates: So do you mean by it a disposition to cause a sensation or experience of red in you?

Tom: Well, I'm not sure.

Socrates: Consider matters, my young friend. Being soluble is said to be a disposition of salt, for example. For if you put salt in water, it dissolves. Being inflammable is a disposition of dry leaves. For if you put a flame to them, they catch fire and burn. Now, is being red like that?

Tom: No: colours aren't like that at all.

Socrates: Good! But *why* are they not like that?

Tom: I s'pose because you can see whether something is red but you can't see whether something is soluble just by looking at it. And looking red to a person is not like dissolving, since you can watch something dissolving, but you can't watch something looking red to another. What you see is the red rose and someone looking at it.

Socrates: Well said. So does 'red' mean the character of an experience?

Tom: No. I don't know what a red, or a green, or a blue experience would be.

Socrates: So you don't think that 'red' is ambiguous?

Tom: No.

Socrates: Now, Professor Swan, don't you think that perhaps it is not so much small wonder that all languages should be ambiguous in respect of names of perceptual qualities, but that it is great wonder that no speaker of any human language should have noticed this ambiguity until Master Locke and his friends came on the scene? Is it not possible that no one noticed any ambiguity because there is none to notice?

Professor Swan: No. I'm not convinced. Asking the student on the Clapham omnibus [*he glares at Tom*] what he thinks about intricate metaphysical matters is, if I may say so, fool ... er ... methodologically unsound. All these unclarities that you have been indicating can be cleared up by a little attention to the semantics of predicates of secondary qualities. Then all ambiguities will vanish. Is it not clear that to be red, for example, just *means* 'to look red to a normal observer under normal observation conditions'.[10] We *define* the colours by reference to subjective colour appearances, that is, by reference to *how things look to an observer* – by reference to his colour experiences. We say that things *are red* if they *look red to a normal observer,* or to put it more precisely, if they have the power to cause normal observers to have the experience of seeing red! So your worries about ambiguity disappear: those roses do indeed look red to a normal observer, so they are red. To be red just *is* to look red to someone with normal vision. And objects look red if they reflect appropriate light onto the retinae of normal observers thus causing a visual experience of red. That clears it all up.

Socrates: Ahhh ... Yes ... I see. But let us not be hasty. You have made a number of very, er ... *precise* ... points, and we must take them one by one.

Let us take up the idea that colours as they are in objects are dispositions to cause experiences of colour in us. Let us first clarify one small point. You have all been talking of colours *as we see them* not being *in* objects. I don't understand you. After all, no one thinks that colours are *in* objects anyway, although the insides of objects may have the same or a different colour from their outsides.

Dr Lockett: We mean only that Colours, Tastes and Smells and such like are not *Entities* in Objects. They are not, as we used to say, *real Accidents.*

Socrates: But why would anyone ever think they are? Colours are properties *of* things, not entities *in* things. They are not thin films on the surfaces of things, and they are not ingredients in things, like flour in bread. No one, other than a confused philosopher or a poor student confused by a confused philosopher, would think that colours are entities, real accidents, *in* things. They may not be what you call 'real accidents', but they really are accidents.

Dr Lockett: Yes, that must be so. I grant you that.

Socrates: Now, you speak of colours *as we see them,* and of sounds *as we hear them,* and so forth. So tell me, how *do* we see colours, hear sounds or smell smells?

Tom: With our eyes, ears and nose – that's how. [*He laughs*]

Professor Swan: [*irritated*] Don't be ridiculous, Tom. You know perfectly well that is not what is meant. To see red is to have a certain subjective experience. Just look at those roses. Now you are having the experience of seeing red. If you were blind or even colour blind, you would not be having that experience. Redness is the qualitative character of your visual experience. It is what some of my younger American colleagues call 'a *quale*' – the 'what-its-likeness of experience', as they put it in their quaint manner. Redness *in objects* [*he glares at Socrates*], on the other hand, is just a causal power to affect observers.

Socrates: Slow down, slow down. Let us try to take things one by one. Of course, Tom, you are right that we see things with our eyes. And surely you agree with Professor Swan that seeing red is very different from seeing blue, although I grant you that this would be misdescribed as a *quality* of the experience. Being enjoyable or horrible are qualities of experiences, but being red or green are not. They are properties of the objects of visual experience, that is: of sight. But let us leave that subject for another occasion.

Apart from Thomas, you all hold that colours, sounds, smells, warmth and cold are no more than dispositions or powers to cause experiences in us.

Professor Swan: Or to be more precise, ascription of a secondary quality to a thing is to be understood in terms of a counterfactual. An object is said to be red if and only if were someone with normal vision to observe it under normal observational conditions, it would look red to him, that is: he would have the experience of seeing red. When we say that secondary qualities, as they are in objects, are dispositions, we mean that objects to which we ascribe such qualities really have no more than a power to cause normal observers in normal observation conditions to have a certain kind of perceptual experience with a certain kind of content.[11]

Socrates: I see. So you think that what you really perceive is a disposition of the object to cause you to have perceptual experiences of certain kinds?

Professor Swan: Yes.

Socrates: We must look closely at the idea of perceiving a disposition or power. Tell me, can one perceive dispositions and powers?

Professor Freiberg: Yeah, sure. One can feel the elasticity of an elastic cord by pulling it, and one can feel the rigidity of a steel rod by trying to bend it. You actualize the disposition by manipulation – that's how you perceive it.

Socrates: But tell me, is *perceiving the actualization of a disposition* perceiving the disposition? Is seeing red, for example, seeing a disposition to cause visual sensations?

Professor Freiberg: Well, … I don't know. But it doesn't matter. Look, you can see that these books can fit into the bookcase, can't you? You just have to see that the bunch of things that you wanna fit into something are smaller than the receptacle they have to fit in. And you can see that a square peg can't fit into a round hole, can't you?

Socrates: I grant you that. But will you grant me that colours (and sounds, and so forth) are not at all like that. For you can't see the disposition of an object to cause you to have what you call 'colour experiences'.

Professor Freiberg: I don't get that, Socrates. I mean, when you look at it, you see its color.

Socrates: Indeed you do, my friend. But not according to your story. The experience of seeing the redness of the wine [*he holds up his empty glass*], as you construe it, is not seeing the power of the wine to cause it to look red to you. The *experience you have*, according to you, is not the power, but the actualization of the power. The power itself is not visible. But now, the experience you have, the experience of the wine's looking red to you, is not visible either, since you cannot see your visual experience. So whichever way you look at it [*he chuckles*], in a manner of speaking, colour is invisible. In short, on your account, you can't see 'red as it is in objects' and you can't see red 'as you see it' either!

Professor Freiberg: Oh! Do have some more [*he pours Socrates another glass*] … That's very interesting. That never occurred to me. I think …

Professor Swan: No, no. This is just naïve. Look here, to see the redness of a red object just *is* to have a visual experience, the content of which has the property – which I called the representational property – of being red, just as feeling the warmth of the sun is to have a tactile experience the representational content of which includes the representational property of being warm. It is absolutely straightforward.

Socrates: I see. Or rather, I don't see *exactly* what you mean. Tell me [*he looks around at all of them*], what do colour words mean?

Dr Lockett: They mean Ideas in the Mind. Simple ideas. Master Locke established quite clearly that simple Names, like 'red' and 'green', and also 'sour' and 'sweet', and 'hot' and 'cold' stand for simple Ideas in the Mind.

Professor Swan: [*brushing Lockett aside*] No, no. We have advanced beyond such primitive notions. I told you: 'red' means to look red to normal observers under normal observational conditions.

Socrates: And what does 'looks red' mean?

Professor Swan: It means that property of the content of a visual experience that you have when you look at these roses or this decanter of wine.

Socrates: And what property is that, pray?

Professor Swan: Come now, Socrates. You are just playing with me. You know perfectly well what property it is – just look at the wine.

Socrates: And when we have drunk it all? How will I then know what it is to look red?

Professor Swan: Well, you'll just have to remember what it looked like, won't you?

Socrates: You mean, recollect the simple idea the word 'red' stands for? Call it up from the storehouse of ideas – my memory. No, no. I forgot – that is too primitive. I shall have to call up a representation of the representative property of the representative content of my having seen the colour of the wine.

Professor Swan: [*annoyed*] I'm not going to bicker over the terminology. You have to recollect the content of the visual experience you had when you looked at the wine.

Socrates: And how am I to know what that so-called content is? Or, as Dr Lockett would say, what idea I had?

Professor Swan: You have to remember it, that's all.

Socrates: But surely, Professor Swan, I have to remember it correctly?

Professor Swan: Yes, of course. That goes without saying.

Socrates: On the contrary, it needs saying again and again. How am I to know whether the mental image, the idea, or the representational content that comes to mind is the right one?

[Long silence]

Professor Freiberg: Say, are you worried about memory scepticism?

Socrates: No, not in the slightest. But we must preserve the distinction between remembering correctly and remembering incorrectly. If I am to know what the word 'red' means, then I must be able to remember *correctly* what mental image or idea goes with it as a defining sample.

Tom: I don't follow.

Socrates: Well, what is the criterion of correctness here? I am supposed to call to mind the idea or mental representation to which I assigned the name 'red', say. So a mental image or idea comes to mind? What shows that it is the right mental image or idea?

Tom: Well, it has to be the same as the one to which you assigned the name 'red'.

Socrates: Of course, but we have not determined what is to count here as the same.

Tom: I don't understand.

Socrates: Look, you can't explain what it is to be seven o'clock on the sun by saying that it is seven o'clock on the sun when it is the same time on the sun as it is here when it is seven o'clock here. On the contrary, we can say that it is the same time on the sun as it is here if it is seven o'clock here and seven o'clock there. But for that, we have first to lay down the determination of times on the sun.[12]

Tom: *[hesitantly]* Yes ... I think I see that.

Socrates: Good. Now does it make sense to use whatever image comes to mind – as a sample that is part of the rule for the use of the colour word 'red', without a criterion that determines what is to count as *the same* idea. Is it just my say-so? Does anything go? After all, if anything goes, then nothing goes! If any image or idea that comes to mind is right, then nothing is right.[13]

Tom: I don't understand, Socrates. I mean, I can remember what red is, and what scarlet or magenta or maroon is.

Socrates: Of course you can, my young friend. But that is because you already *have the concepts* of red and of shades of red, like scarlet, magenta and maroon. You already know what these words mean.

Tom: Yes, of course. But ... I ... can I ask something silly?

Socrates: That is the very best thing to ask.

Tom: Well, I mean ... how do I know?

Socrates: Well, how would you explain what 'red' means to a child?

Tom: Well, I'd point at something red, and I'd say 'That colour is red'

Socrates: Exactly. Indeed, isn't this part of what it is to know what red is? Is this not a criterion for your knowing what the word means? And what you point at is a public sample, visible to all normal sighted people – not at a private mental sample that only you have and only you can apprehend.

Tom: Yes, I see. Of course.

Professor Swan: [*angrily addressing Socrates*] You've been talking to that damned fellow Wittgenstein. That's his private language argument. I've never been able to understand what he was going on about. Just a weird Austrian – all charisma and no content. Just waffle.

Socrates: Yes, it is his argument – and a very fine argument too. But I can't explain it all to you now, as I must go soon. I am going to meet him for tea. But I hope that you can see that not only is 'red' not the name of a power or disposition of red objects, it is not the name of an idea or of a property of a representational content either. It does not have an objective and also a subjective meaning. It is not ambiguous at all. The truth of the matter is very simple, if only you would take off your blinkers, it is …

Tom: [*laughing*] The name of a colour!!

Socrates: [*chuckling*] Quite so. We explain what the colour red is, and so too what the word 'red' means, by pointing at a sample in good light, explaining: that ☞ colour is red. That is a rule for the use of the word, for anything that is *that ☞ colour* is rightly said to be red. And we can determine whether something is red – or more probably whether it is magenta or maroon – by comparing the sample with the object, holding it up against the object to see whether the object is the colour of the sample. Normal observation conditions are those conditions under which coloured things are visibly the colour they are. Normal observers are observers who can discriminate coloured things under normal observation conditions. We teach colour words *under* normal conditions, but not *by reference to* normal conditions, just as we teach the use of colour words *to* normal observers, but not *by reference to* normal observers.

Professor Freiberg: Neat.

Socrates: So is it not obvious now that were colour words names of what you called private sensations in the mind or brain, we could never understand them. There could be no such names at all. You have granted that looking red is not the name of a sensation in the mind or brain. Now you must surely grant that we understand the expression 'looking red' only insofar as we already know what 'being red' means. For something looks red only if it looks as red things do.

And we explain what it is for something to be red by reference to public samples visible to all who enjoy normal eyesight.

Dr Lockett: Most ingenious, most ingenious.

Socrates: Good. Perhaps we are making a little progress. Now, you said something curious …

Tom: Socrates, I have a question.

Socrates: Go ahead, my dear boy.

Tom: Well, just before you joined us, Ronnie was explaining that science can explain why we see red, green or yellow things, or why we feel warmth or cold, without even mentioning secondary qualities, but only wavelengths of light, and so forth. And he was suggesting that this shows that secondary qualities are not real, but only apparent.

Socrates: I see. [*He turns to Professor Freiberg*] So you want an explanation of why, when we look at these splendid red roses, we see red roses rather than white daisies or [*he chuckles*] pink elephants?

Professor Freiberg: [*laughs*] No, no, we can leave the pink elephants out. No, the point is that we wanna explain our perceptions. Now, in understanding, even roughly, why things appear colored to us, we don't even have to refer to colors, only to light and reflectance coefficients. So we can abandon the whole idea that things *really* have one color or another. The hypothesis that objects have intrinsic colors, smells, tastes, thermal qualities and so on, in addition to their primary qualities, provides a poorer explanation of why they appear to have colors, smells, tastes and are warm or cold than the hypothesis that the primary qualities of objects and their effects on our nerve endings are responsible for their appearances.[14]

Socrates: How interesting. Now, since I am ignorant in these matters, explain to me why you want an explanation of why, when we now look at these red roses, we see red roses? What else might we have seen?

Professor Freiberg: Well, I guess we might seem to see yellow roses.

Socrates: Well, *then* you would have some explaining to do. But what is there to explain about someone's seeing red roses when he looks at red roses in normal circumstances?

Professor Freiberg: Look, Galileo wrote that he couldn't believe that there exists in external bodies anything other than their size, shape or motion … which could excite in us our tastes, sounds and odors – and our visual experiences of seeing colors, and our auditory experiences of hearing sounds. We know now that we need more than just shape, size and motion, but still, he was on the right track.

Socrates: Was he? Was he really? But if you are trying to explain why, when I look at the red roses, I have what you call 'the experience of something's looking red to me' then you have gone wrong before you have even started, my dear fellow.

Dr Lockett: Allow me, Socrates – but I don't see why.

Socrates: Well, you agreed that for something to look red to a person, it must look to him to be what red things are, namely red.

Dr Lockett: I hesitate … Yes, yes.

Socrates: So looking red to a person presupposes the intelligibility of things' being red.

Dr Lockett: Hmm. I hesitate … Yes, I suppose so.

Socrates: So in explaining why things appear to have the colours they do, you presuppose that, for the most part, they do have the colours they do.

Dr Lockett: But in truth, Socrates, we wish to understand why things are the Colours they are, what makes them appear red or green, yellow or blue.

Socrates: Well, Tom, what makes post-boxes in England red?

Tom: Well, I guess because they're painted in red paint – to make them salient.

Socrates: Quite so. And what makes red paint red?

Tom: Well, I guess it is the red pigment in the paint.

Socrates: And what makes grass green?

Tom: Well, in one sense, water and sunlight – otherwise it turns yellow; and I guess, in another sense, because it contains chlorophyll, and chlorophyll is green.

Professor Freiberg: But surely, Socrates, what makes red things red is the fact that their surface structure absorbs all the light and reflects only light in the waveband between 700 and 635 nanometers. And what makes green things green is that the only light they reflect lies in the 530–470 nanometre range.

Socrates: Well, my friend, you are an authority on such matters, about which I know little. But tell me, when a mason carves a sphere, what makes the sphere spherical?

Professor Freiberg: Obviously the mason does, with his chisel.

Socrates: Indeed. And when a builder constructs a well-dressed wall one *kalamos* thick, what makes it solid?

Professor Freiberg: I guess he packs the gap between the dressed stones that make up the two surfaces of the wall very tightly with earth and rubble.

Socrates: Quite so. But is it not the atomic structure of the surfaces that makes the wall solid?

Professor Freiberg: Ah! Yeah. I see. Yeah, that is what makes all solid things solid in one sense ... Well, in another sense – no. I mean, that is not what makes the solid wall solid. It is, as a matter of empirical fact, what it *is* for any chunk of stuff to be solid. Solid objects don't flow to take on the shape of their container, like liquids, and they don't expand to fill the available space, like gasses. They're structurally rigid and resistant to change of shape and volume. The ions, atoms or molecules of solid things are tightly bonded together either into polycrystalline structures, as we find in metals, or amorphously. This tight bonding explains what it is for something to be solid and also explains solidity. But, of course, you can have a wall built of stone, that is – of stuff that is solid – that is so badly built that the wall is not solid at all, but easily demolished.

Socrates: Fascinating. But now, does this explanation show that solid things are not really solid.

Professor Freiberg: No, of course not. It explains the constitution of solid material. I mean, it explains, in one sense, what it is for material to be solid, as opposed to liquid or gaseous. It doesn't explain what it is for something to be solid as opposed to being friable, hollow or porous.

Socrates: Quite so. The crystalline or amorphous arrangement of the atoms of solid things does not explain solidity away. It does not even explain what makes a solid thing solid, in the sense in which your explanation of the solid wall explains what makes *it* solid. It explains what it is empirically for anything to be solid as opposed to being liquid or gaseous, does it not? After all, it was an empirical discovery that the atoms of which solid things consist are arranged in lattice-like structures, wasn't it?

Professor Freiberg: Yeah. I guess that's right. OK.

Socrates: So why do you think that your explanation that all red things reflect light waves of such-and-such a wavelength explains colour *away*? It explains what it constitutively is, as a matter of fact, for a surface to be red – not what makes it red, in the sense in which a coat of red paint makes something red, or staining cloth with cochineal makes it red.

Professor Freiberg: Yeah. I see. So there are quite different notions of explanation at work here.

Socrates: Good. We're slowly getting there! But now, there was one more point that we have to sort out together. I understand that you were saying that science has no *need* to invoke colours or sounds, heat or cold, in its explanations.

Professor Freiberg: Yeah, that's right.

Dr Lockett: Yes, yes. That is so.

Socrates: Now tell me, if it is true that science has no need to invoke colours, why would that show that there are no objective colours.

Professor Swan: Well, we need not postulate colours, any more than we need postulate sounds, smells, tastes, heat and cold.[15]

Socrates: How curious. Tell me then, do you need to postulate me?

Professor Freiberg: What do you mean? I can see you, talk to you, touch you.

Socrates: And you can't smell the fragrance of the roses?

Professor Freiberg: That's different.

Socrates: Why is that different? It is not a hypothesis that you are talking to me, and no more is it a hypothesis that this wine is red, and that the roses are fragrant. If you see empty glasses and an empty wine bottle, you may form the hypothesis that people have been drinking wine. But if you see us drinking wine, that's not a hypothesis – it's seeing us drinking. It's what confirms the hypothesis.

Professor Freiberg: Yeah, I guess that's right.

Socrates: Indeed. There is no such thing as postulating what is evident. You need but look to see.

Professor Freiberg: OK. But the fact is that secondary qualities do not occur in a scientific description of the world, and play no role in explanations of phenomena in physics.

Socrates: But then you too play no role in a so-called scientific description of the world. Does that mean that *you* are not real or are 'merely subjective'? Or, haha, a postulate!

Professor Freiberg: I didn't follow you then, Socrates.

Socrates: Well, let me try again. Correct me if I am mistaken, but I take it that what you call the scientific description of the world is a description given by physics.

Professor Freiberg: Yeah, I guess so.

Socrates: Does physics describe mountains and rivers?

Professor Freiberg: No, of course not, that's the job of geography.

Socrates: But that does not banish rivers and mountains from existence, does it? Now does a scientific description of things include the history of peoples?

Professor Freiberg: No, physics isn't history.

Socrates: Quite so. But it is still true that we won at Marathon and the Persians were defeated, is it not? And does a scientific description include the laws of peoples.

Professor Freiberg: No, of course not. That's for jurisprudence.

Socrates: But you don't think that the histories and laws of peoples do not exist just because physics has no interest in them, do you?

Professor Freiberg: No, of course not.

Socrates: Then tell me why you think that secondary qualities don't exist just because you suppose that they don't occur in a scientific description of the world?

Professor Freiberg: Yea. I see.

Professor Swan: No, no, Socrates. Now *you* are going too fast. Secondary qualities are causally impotent save for their effects on sentient creatures. So if we want to explain *physical* phenomena, our description will not include any mention of the secondary qualities of things. For they can't explain anything that happens outside the domain of biology and human affairs.

Socrates: Do you mean that I whitewashed my house every few years for nothing?

Professor Swan: What do you mean?

Socrates: Well, did we not learn that black things warm up more quickly than white things? Do we not wear white clothes rather than black ones because they are cooler in the summer? And is that not because white surfaces reflect rather than absorb the heat of the sun? Were I to paint my house black, the colour of the house would make it intolerably hot in the summer.

Professor Swan: No, no. That is not a consequence of its being black. It is because being black correlates with having a tendency to heat up more, and this correlation exists because the colour and the tendency have the same cause, namely the absorption of light.[16]

Professor Freiberg: Hey, wait a minute. What makes Socrates's house white is that it is whitewashed. If he hadn't whitewashed it, it wouldn't have been white. The color makes all the difference to its being cool in the summer heat. And what makes a lump of wax melt in the sunshine is the heat of the sun, and what makes the lemonade we're gonna drink cold is that we put ice into it, and the ice is cold. You don't wanna to deny the obvious.

Professor Swan: No, no, colours are causally impotent. What makes things black or white, red or green and so forth is the specular reflectance properties of their surfaces, which is determined by the molecular structure of the surfaces, as you know full well, Professor Freiberg.

Professor Freiberg: Aw, come on. We can explain why black surfaces absorb all the light that impacts them, and why white surfaces reflect all the light that impacts them, but that doesn't mean that the color of a white house doesn't explain why it heats up less than a black house.

I mean, y'might as well say that a ball doesn't bounce off the surface of the wall because the wall is solid, but only because the molecules of the wall are arranged in a stable lattice structure common to all solids. Y' know, to explain isn't to explain away. The fact that heat can be explained as mean kinetic energy doesn't mean that the heat of the sun isn't the cause of the wax's melting.

Socrates: Well, we can leave that one for the two of you to sort out. But surely you must grant that whether or not being of such-and-such a colour can make physical things happen, whether or not possession of a secondary quality can cause physical events, is irrelevant to the question of whether secondary qualities are objective properties of objects around us.

Professor Swan: All right. Whether being coloured, hot or cold, wet or dry may, or may not, feature in a causal explanation of events in the physical world does not, as such, show that they are not objective. But surely they are relative to human sensibility. The blind can see no colours at all; the colour blind cannot see the difference between red and green; bees apparently can see ultra-violet spots on certain plants, which we cannot see. Does a dog-whistle make a sound? Not one that we can hear. So secondary qualities are all mere appearances relative to us. They are relational properties, not absolute ones.

Socrates: Ah! Now that is a large subject. And one which I am afraid we must leave for another day. Now, I am sorry I must leave you, even though there are so many things yet to be settled. But I must not be late for tea. Still, perhaps you can sort them out by yourselves. It just needs a little thought, you know. [*Rising from his seat, he turns to Tom*] Thomas, I should like you to accompany me. I want you to meet my friend Wittgenstein.

Tom: [*aghast*] Oh! Oh, Lord! [*Tom rises too. Socrates puts his hand over Tom's shoulders*]

Socrates: Come now, don't be afraid. Just be you natural self, as you always are, and Ludwig will take to you. Good day, gentlemen. [*They walk off together*]

Dr Lockett: Well, I must think upon all this. A remarkable conversation!

Professor Swan: No. No. Not really. The old boy really hasn't kept up with the literature, you know. And he's been talking to Witters to boot. No substance, really. No rigour or precision.

Professor Freiberg: Well, ... I don't know. I thought he was pretty nifty on his feet. And he sure gave us something to think about.

Notes

1   Galileo, *The Assayer*, p. 28.
2   Newton, *Optics*, 4th ed. p. 109.
3   Eric Kandel, James Schwartz and Thomas Jessell, *Elements of Neural Science and Behaviour* (Appleton and Lange, Stamford, CT, 1995), p. 370.
4   S. Zeki, 'Colour Coding in the Cerebral Cortex', *Neuroscience* 9 (1983), p. 764.
5   Christopher Peacocke, *Sense and Content: Experience, Thought, and Their Relations* (Clarendon Press, Oxford, 1983), p. 5.
6   See Wittgenstein, 'Philosophy of Psychology – a Fragment', §55, in *Philosophical Investigations*, 4th ed. (Wiley-Blackwell, Oxford, 2009),
7   Peacocke, *Sense and Content*, p. 7.
8   G. Frege, *The Foundations of Arithmetic* (1884), trs. J. L. Austin (Blackwell, Oxford, 1959), §26.
9   J. L. Mackie, *Problems from Locke* (Clarendon Press, Oxford, 1976), p. 11.
10   A common view among twentieth-century philosophers, e.g., B. Russell, *The Problems of Philosophy* [1912] (Oxford University Press, London, 1967), p. 2; A. J. Ayer, *The Central Questions of Philosophy* (Weidenfeld and Nicholson, London, 1973), p. 83; G. Evans, 'Things without the Mind', in Z. van Straaten (ed.) *Philosophical Subjects, Essays Presented to P. F. Strawson* (Clarendon Press, Oxford, 1980), p. 98; C. McGinn, *The Subjective View* (Clarendon Press, Oxford, 1983), p. 5.
11   A view advanced by numerous twentieth-century analytic philosophers, e.g., J. Bennett, *Locke, Berkeley, Hume: Central Themes* (Clarendon Press, Oxford, 1971); Evans, 'Things without the Mind', p. 94; J. McDowell, 'Values and Secondary Qualities' in T. Honderich (ed.), *Morality and Objectivity, A Tribute to J. L. Mackie* (Routledge and Kegan Paul, London, 1985), pp. 111–12.
12   The example is derived from Wittgenstein, *Philosophical Investigations*, §§350–51, but put to a slightly different use.
13   Wittgenstein, *Philosophical Investigations*, §258.
14   Bernard Williams, *Descartes: The Project of Pure Enquiry* (Penguin, Harmondsworth, 1978), p. 242; Thomas Nagel, *The View from Nowhere* (Oxford University Press, Oxford, 1986), p. 76.
15   J. L. Mackie, *Problems from Locke*, pp. 18–19.
16   See J. Hyman, *The Objective Eye* (University of Chicago Press, Chicago, 2006), pp. 19–20.

SUPPLEMENTARY READING

P. M. S. Hacker, *Appearance and Reality* (Blackwell, Oxford, 1987).
Barry Stroud, *The Quest for Reality* (Oxford University Press, Oxford, 2000).

SECTION 4

TWO DIALOGUES ON THOUGHT

INTRODUCTION

There can be few thinking people who have not, at some time or other, wondered about the nature of thought. Thought seems to be able to do so many remarkable things. Sitting at home, one can think of someone on the other side of the world, and one's thought, it seems, unerringly reaches right out to him. It transfixes him, as it were, with an unerring mental arrow – *this* → is the person of whom I am thinking, and no other. How can this be? Indeed one can think of people who no longer exist, of Solon or Socrates. How does one's thought reach back into the depths of the past to pinpoint just Solon and not Solomon, or just Socrates and not Hippocrates? Moreover, can one not think of personages who *never* existed, such as Adam and Eve, Zeus or Aphrodite? How is this magic effected?

Equally baffling – once one thinks of it – is the sheer speed of thought. Faced with an urgent practical predicament, one often comes to a conclusion like greased lightning, although after the event it may take one ten minutes to explain the reasons why one decided to do what one did. One may see the solution to a theoretical problem in a flash, but showing it to be the solution may take one half an hour. The great Cambridge mathematician G. H. Hardy, in the course of a lecture, is said to have written an equation on the blackboard and to have remarked casually: 'That's self-evident'. A student in the front row exclaimed, 'It doesn't seem self-evident to me, Professor Hardy'. Hardy picked up a piece of chalk and started writing formulae on the blackboard. Half an hour later, having covered three blackboards with figures, he underlined the final line of the proof and turned triumphantly to the student exclaiming 'I told you it was self-evident!' How can this be? How can one think one's way to the solution of a complex proof in a flash, when it takes half an hour to spell out what one thought? Thought is indeed a mysterious process.

Well, it may be mysterious, but is thinking really a process? Of course, it seems to be. It usually takes time, as do activities. One may be engaged in thinking, just as one may be engaged in digging the garden. One may be interrupted in the middle of thinking, just as one may be interrupted in the middle of speaking. Is it not *obvious* that thinking is an activity of the mind? No, on reflection, not at all. For what does one *do* when one is thinking? Maybe one talks to oneself. But talking to oneself is not thinking. One can think without talking to oneself, and talk to oneself without thinking. Maybe various mental images cross one's mind while one is thinking. But having an image before one's mind's eye is not thinking, and one can think without having any mental images before one's mind. Thinking seems to become more and more intangible, more and more insubstantial.

Moreover, is thinking an activity *of the mind*? I may think something over, but does my mind think it over? And if so, how does it tell me what conclusion it has come to? If it informs me what the results of its activity of thinking are, in what language does it inform me? Does my mind speak different languages? Or is it that I think *with my mind*? And how do I do that? Do I think with my mind in the manner in which I walk with my legs? Is my mind the organ of thinking as the legs are the organ of locomotion? Or do I think with my brain? But how do I do so? I can move my legs to walk or run, but I cannot do anything with my brain. Or is it my brain that thinks, and then informs me what it has thought. But to do so, it must surely think in a language! What language does my brain speak? And do I always understand what it says?

Must one really think *in a language*? Surely, when I think, I must think *in* something – just as when I speak, I must speak *in* some language or other. Just as speech requires a medium, so too, it seems, must thinking. One may think in language, or one may think in mental images or ideas, but one must think *in* something. Nevertheless, one should pause and ask *why* one must. What would happen if one didn't? Admittedly we talk to ourselves in the imagination a great deal, but talking to oneself in the imagination is not the same as thinking. And when we speak with thought, we surely do not talk to ourselves in the imagination at the same time! So what *is* the thinking that is involved in thoughtful speech?

Moreover, is language really necessary for thinking? Do small children not think before they have learnt to speak? And surely the higher animals are capable of thinking, even though they cannot speak. They can solve quite complex problems, some of them make and use tools. They can recognize themselves in a mirror, and surely that betokens self-consciousness. But isn't self-consciousness itself a form of thought?

The thoughts of others, we are inclined to say, are hidden from us. 'I wish I knew what is going on in his head', we sometimes remark. But are thoughts in our heads? Is that where thoughts are stored? Or is that where they occur? We do say that a thought flashed through our mind, or that we have a thought at the back of our mind. But is the mind in the head? It is the brain that is in the head. But is the brain the same as the mind? Does the mind then weigh three pounds? How on earth could a thought be in the brain? If we ask where

Archimedes had his famous Eureka thought, the answer is 'In his bath', not 'In his brain, of course'.

No matter where they occur, it is surely a matter of indisputable fact that our own (conscious) thoughts are patent to us. The thoughts of others are hidden in their head. I cannot *really* know, with utter certainty, what others are thinking. There is always room for doubt. I may have made a mistake. But is it not strange that my own thoughts, or at any rate my own conscious thoughts, are as it were transparent. If I think something to be so, then I know that I so think. I cannot be mistaken. I am certain, and I cannot doubt that I so think. So it seems as if, in the world of my own thoughts, I am like God – omniscient and infallible. Is that not a mystery?

These are the great themes that are debated in the following two dialogues. They take place in Elysium, where a jovial but penetrating Socrates guides the discussion, presses the disputants and ruthlessly cuts out the nonsense by his questions. John Locke presents the idealist conception of thought characteristic of the seventeenth century and of subsequent British empiricism and early psychology. Frank, an American neuroscientist, presents the scientific views of the last few decades. Paul is an Oxford analytic philosopher of the 1950s, skilled in linguistic analysis. Alan is a Scottish post-doctoral student. The first dialogue breaks off for dinner. The second post-prandial dialogue is focused upon the relationship between thought and language.

Sixth Dialogue

THOUGHT

Protagonists:

Socrates: gravelly voice, slight regional accent. Dressed in ancient Greek manner. (It should be noted that Socrates has spent a great deal of time talking with his friend Wittgenstein.)

Paul: a middle-aged Oxford don of the 1950s, dressed in well-cut sports jacket, waistcoat and tie, Oxford English accent.

Alan: a bright Scottish post-doc, tweed jacket and woollen tie, soft Scots accent.

Frank: a contemporary American neuroscientist in his forties, casually dressed, Californian accent.

John Locke: in seventeenth-century scholar's garb. A pedantic and slightly reedy voice.

The scene is a garden in Elysium. The evening sun is shining. A rich verdant lawn is surrounded by flower beds and rose bushes in bloom, with tall trees behind. Beyond, there is a beautiful view of lake and mountains. Frank, Paul, Alan and John Locke are seated on comfortable garden chairs, placed in the shade beneath some trees. There is a low table on which there are some scattered books, a couple of bottles of wine and spare glasses. Each person has a glass of wine before him from which he occasionally drinks. Socrates has just joined them, and is about to sit down.

Socrates: Thank you, I should be delighted to join you. And I fancy that I might assist you in consuming one of those excellent bottles of Nectarian. Haha! Drinking gets my approval, my friends. Is it not indisputable that wine refreshes the soul, allays our worries and feeds the flame of good cheer? [*Paul pours him a large glass of wine*] Of course, in moderation. [*He quaffs half his drink and smacks his lips*] Ah, excellent! So, what is it you were talking about?

Paul: We've been talking about thinking, Socrates, and the longer we talk about it, the more mysterious *it* becomes, and the more confused *we* become. So we thought that you might be able to help us.

Socrates: Don't you know what thinking is? After all, you are as familiar with thinking as you are with walking, and you know perfectly well what walking is.

Alan: Ay. I can walk – and think too. I ken wha' walking is and wha' thinking is. But while if you ask me wha' walking is, I can easily tell you, if you ask me wha' thinking is, I find it very difficult to say. I mean, we've been talking about it for the last hour, but we always seem to find ourselves up a dead-end.

Socrates: What sort of dead-end?

Alan: We agree tha' thinking is *an activity* tha' people engage in. And we agree that it's an *inner* – private – activity. For we ken our own thoughts in a way in which others canna ken them. But while it seems to Mr Locke and to me that thinking's an inner activity o' the mind, Frank thinks it is an inner activity o' the brain. And we canna seem to resolve this.

Frank: Yeah, that's right. We *know* that we think with our brains. Actually, we can be more specific, we think with the frontal cortices – that's where thinking takes place. If you put someone under an fMRI scanner – [*he turns to Socrates*] that's a modern machine that enables us to look into the brain and observe regional variations in its activity – then you can *see* thinking occurring.[1] So thinking just is a neural activity of the human brain.

Alan: But that canna be right, Socrates. After all, we ken wha' we think. But if thinking were an activity o' the brain, then we wouldna ken wha' we think, since we dinna have a clue wha's going on in our brains.

Locke: Methinks Master Alan is in the right. Every Man is conscious to himself, That he thinks, and that which his Mind is employ'd about whilst thinking are the Ideas with which his Mind is furnished. For do we not perceive the Operations of our own Minds within us? 'Tis plain we apprehend Thinking by Inner Sense, or by Reflection – by which I understand that notice the Mind takes of its own Operations. 'Tis this that Master Leibniz called Apperception,[2] by which he intended what I denominated the Perception of our Perceptions, and 'tis this that was denominated Introspection by Sir Matthew Hale.[3] But no Man is conscious to himself of the Operations of his own Brain.

Frank: But that's no argument against the identification of thinking with neural activity. Most of the operations of the brain are *unconscious*.[4] But some are *presented* to consciousness, and it is these that we perceive by introspection. When we're conscious of thinking, we see the operation of the brain *from the inside*. When we see the neural activity on a scanner, we see thinking *from the outside*.

Locke: This I cannot acknowledge. 'Tis true that these Subjects have in all Ages exercised the learned part of the World with Questions and Difficulties. But to say thus that we can see Thinking from the Inside and from the Outside is a confused Notion taken up to serve an Hypothesis. For the most that can be said of it is this: That when a Man thinks, one can see, in this ingenious Machine of which you speak, what occurs in the Man's Brain.

Paul: And anyway, it's quite wrong to say that *most* of the activities of the brain are unconscious. They're *all* unconscious ... Or better, the activities of the brain are neither conscious *nor* unconscious – just like the activities of the kidneys or liver. What is unconscious *could be* conscious, but isn't. Neural processes in my brain are not processes that I *could* become and then be conscious of. They are non-conscious.

Frank: No. I don't see that at all. No, no, I think ... . [*He is interrupted by Socrates*]

Socrates: You all seem to be trying to weave a tapestry before you have removed the knots from the thread. Small wonder that there are snags everywhere.

Let me see. You all agree that thinking is an activity. You all agree that it is, so to say, 'inner' – something that is private to the thinker. And you all agree that we know what we think by introspection or inner sense. But you quarrel over whether it is the mind or the brain that thinks. Oh, and Master Locke holds that thinking is an operation with ideas. Are you all ready to go along with that?

Paul: Well, no. I'm not. Thinking is not watching a parade of ideas in the mind. Surely, thinking is operating with words.[5] We think in words and sentences. And didn't you yourself, Socrates, say that thinking is a discourse the mind carries on with itself about any subject it's considering?

Socrates: Did I say that? I don't think I did, you know. Who told you?

Paul: Why Plato did. He put those very words into your mouth in his dialogue the *Theaetetus*,[6] and in the *Sophist* he has you say that when the mind is thinking, it's simply talking to itself, asking questions and answering them, and saying yes and no.[7]

Socrates: Did he really! The naughty fellow [*he chuckles*]. No, no, I said no such thing. After all, it should be obvious that we can say things to ourselves in the imagination *without thinking*, as when we recite a poem to ourselves. After all, when I recite Ajax's famous lines to myself, I don't *think* that.

> Here I am, the bold, the valiant,
> Unflinching in the shock of war,
> A terrible threat to unsuspecting beasts.

I'm no Ajax, but I know my Sophocles.[8] [*He laughs*] And when we want to fall asleep, we count goats in our imagination – we say to ourselves: 'One goat, two goats, three goats' and so on, *in order to stop ourselves from thinking*. So we can obviously talk to ourselves without thinking. But equally, we can think without talking to ourselves, as when we speak with thought – as I am now doing [*he laughs again*], and equally, when we engage in practical activities with thought and concentration – as when we carve a beautiful sculpture with the skill of a Praxiteles, or paint a painting with the subtle thoughtfulness of a Zeuxis.

Locke: I would be glad also to learn from these Men, who so confidently pronounce that thinking be inner Discourse, how they would explain that the Words of such inner Discourse have Signification. For Words are themselves signs of internal Conceptions. If they stand not for Ideas, they are Signs of nothing, and are without Signification. It is Ideas that are the materials for Thought.[9]

Frank: Hey, hold it. Did I hear you right? Y'mean to say that we think in ideas and then translate our thoughts into words?

Locke: This I cannot forbear to acknowledge. Did Master Hobbes not write that the most noble and profitable Invention of all other was that of *Speech*, consisting of *Names* or *Appellations*, and their Connection; whereby Men register their Thoughts and also declare them to one another? The general Use of Speech, he said, is to transfer our mental Discourse into verbal, the Train of our Thoughts into a Train of Words.[10] Besides this, he wrote also that Men, having invented Language to the end of showing others the Knowledge, Opinions and Conceptions which are within them, have by that means transferred all that Discursion of their Mind, by the Motion of their Tongues, into Discourse of Words, and *Ratio* now is but *Oratio*.[11] I would have you know that the Commonwealth of Learning in my times was not without other Master Builders. Did not the admirable Arnauld, in his lasting Monument *The Art of Thinking*, write that Words are Sounds distinct and articulate, which Men have taken as signs to express what passes in their Mind?[12]

Frank: OK. That's what they wrote. But what do *you* think?

Locke: I do hold what I wrote. Man, though he have a great variety of Thoughts, and such, from which others might receive Profit and Delight; yet they are all within his own Breast, invisible, and hidden from others. The Comfort, and Advantage of Society, not being to be had without Communication of Thoughts, it was necessary, that Man should find out some external sensible Signs, whereby those invisible *Ideas*, which his Thoughts are made up of, might be made known to others. Thus we may conceive how *Words* come to be made use of by Men, as *the Signs of* their *Ideas*.[13]

Frank: Well, I don't go along with that. It seems to me that words are, for the most part, names of things, of objects and of properties and relations, and so on.

Socrates: Well, my dear fellow, let's defer that subject for the moment. We surely have enough on our plate already. Let's start with the idea that thinking is an activity.

Alan: But we all agree wi' tha'.

Socrates: Precisely. When men argue for centuries over some great subject in philosophy, like the nature of thinking, we are all tempted to examine the matters they disagree over, and after carefully attending to both sides of the debate, we try to decide between the contestants.

Frank: Yeah, sure we do. How else should we get at the truth?

Socrates: Well, now, if a philosophical debate has been going on for century after century without resolution, like the debate about the nature of thought, then something must be wrong. For it is not as if we are dealing with an unfamiliar phenomenon. And it is not as if our puzzles and muddles about thinking can be resolved by observation or experiment. And it's not as if the question is obviously so much *more* difficult than questions about the nature of the cosmos. But much more progress has been made in the investigations of the cosmos.

Alan: So what are we t' do?

Socrates: We should look closely at what is *agreed upon* by all sides, and challenge precisely those propositions. [*He chuckles*] In philosophy, we go wrong before the first step is taken. We go wrong in something we all take for granted.

Locke: That is most ingenious.

Socrates: So let's proceed to examine the idea that thinking is an activity.

Paul: Or process. Thinking is something we *can* do voluntarily, intentionally and deliberately. And that surely is an activity. But very often we *find* ourselves thinking about this or that, and thoughts cross our mind whether we like it or not. That sort of thinking is not voluntary, and I suppose that we should conceive of it as a mental process rather than activity.

Socrates: All right, Paul. But I don't think that matters for present purposes. Now what, in your view, speaks for the idea that thinking is an activity?

Locke: Do we not busy ourselves with thinking? Does Thought not take time, no less than Motion? And can we not often remove our Contemplation from one Idea to another at pleasure? Such phaenomena speak for the Supposition that Thinking be an Activity of the Mind.

Alan: Tha's right. And we can think quickly, or take our time and think slowly. Thinking may be easy or difficult, just like climbing – which is an activity. It can be interrupted, as when the telephone rings while one is thinking one's way through a problem. And thinking, once broken off, can be resumed, as when the telephone conversation stops and one can turn back to the problem.

Paul: Indeed. And one may be wholly absorbed in thinking. The utterance 'I was thinking' answers the question 'What were you doing all that time?' We say 'Don't disturb him, he is thinking his way through a difficult problem' – and that surely shows that thinking is an activity.[14]

Frank: Yeah. And whether your pupil Plato was right or wrong about thinking being a silent conversation with oneself, it's undeniable that we often-times talk to ourselves in the imagination. And talking to oneself in the imagination, just like talking to someone else out loud, is an activity. I don't see how anyone can deny this, and I don't see why we should challenge it.

Socrates: Good! And now let's look more closely. Remember that we are speaking, at any rate for the moment, of only one aspect of thought. We are not concerned with thinking *qua* opining, as when we say that someone thinks that this statue is more beautiful than that one. Nor are we discussing thinking *qua* qualified judgement, as when someone says that he thinks, but is not sure, that Speusippus will attend Plato's lecture tomorrow evening. Our interest is in reasoning, reflecting, deliberating, ruminating and musing, in thinking before we speak, in speaking with thought, in doing something with thought and attention. Here it does indeed look as if thinking is an activity.

Paul: There is a further point, Socrates. One can strengthen our case that thinking is an activity – if indeed it needs strengthening at all – by an appeal to grammar. The verb 'to think', like the activity-verbs 'to speak', 'to work' and 'to talk', has a continuous tense – we say such things as 'Don't interrupt me, I'm thinking'. It has an imperative mood, for we can say to another 'Think it over!'. Like all activities, it can be qualified by manner adverbs, such as 'quickly', 'laboriously', 'reluctantly'. And similarly it can take 'for ... sake' constructions, as in 'I thought quickly for Jill's sake', which is like 'I worked quickly for Jack's sake'. These features, according to the grammarians, are marks of activity-verbs. So how can there be any objection to something as evident as the claim that thinking is an activity – in my view, an activity of the mind.

Socrates: Very good. So we can say that *thus far*, and *in these respects*, thinking is an activity.[15]

Frank: OK. So where's the snag?

Socrates: Look at the kinds of things we *call* 'activities', Frank. Do they not consist of a sequence of acts unified by some end? They may be a repetition of acts, like digging; or an ordered sequence of acts, like cooking according to a recipe; or they may be an unordered but non-repetitive sequence of acts determined by circumstances, as when we engage in conversation with our neighbours. Such sequences of acts take time, can be interrupted and later resumed, and at any given time in the course of the activity, one is doing something that is a phase or stage of the activity. So far, so good?

{ **Locke:** 'Tis undeniable. Pray proceed.
{ **Frank:** Sure.

[*The others nod*]

Socrates: Now look at thinking, musing and ruminating. One may be thinking how to solve a geometrical problem or an empirical problem in which we reason from evidence to conclusion. Or one may be doing something with thought. One may be speaking thoughtfully. One may be concentrating as hard as one can on the task at hand. But at any given moment, *there may be nothing at all going through one's mind* – no mental images, and no internal dialogue with oneself. There may be no mental events or acts occurring that are phases or stages in a process or activity. But that does not mean that one has stopped thinking or is not thinking. And even if something is crossing one's mind, what is going through one's mind may be no more than saying to oneself 'Hm!', 'No' or 'Aha!' – and that is surely not *what* one is thinking.

Frank: But Socrates, that's just because it's your brain that is thinking, and a large part of what it thinks, when it's solving the problem you face, is unconscious. Thinking is going on alright, but it's not conscious thinking.

Paul: And I fancy that cognitive scientists would claim that most of the activities of the *mind* are unconscious.[16] Even if nothing is going on in the conscious mind at any given moment, a whole host of procedures of protocol analysis, information processing, computation and parallel distributed processing, and so forth are going on in the unconscious mind.

Alan: No, no! Tha's just a fiction. It's …

Socrates: [*interrupting*] We'll come to these points later, my friends. We can't tackle all the problems at the same time. Every worthwhile philosophical confusion is held in place by many different struts. Each such strut has to be examined separately as one walks around the problem. And each one has to be condemned for good reasons. Only then can we bring down the artifice of confusion. But we can't examine all the supporting struts simultaneously.

Let me move on to a second objection to the idea that thinking is an activity – just like physical activities, only an 'inner', mental, one. We *can say* what we're thinking or have thought. But we can't say what we're thinking *on the basis* of what passes in our mind *when* we are thinking. We cannot read off what we're thinking from any inner goings-on in the mind. But if thinking were an inner activity, we ought be able to do so.

Locke: I should be in your Debt, Socrates, were you to explain this a little more particularly, for I see it not.

Socrates: Of course. Let's suppose that while you're thinking, you say to yourself 'I'm sure he'll be there on time'. Now, you can't read off from this fragment of inner speech the fact that what you have concluded is that Crito will join our symposium at Agathon's mansion at sunset tomorrow. By contrast, if you see me digging in the field, planting seeds and covering them with soil, watering the soil, and so on, you *can* read off in words a description of what I am doing.

Paul: But if you say out loud 'I'm sure he'll be there on time', we can't read the thought expressed off the overt utterance alone either.

Socrates: No, of course not. But there are public reference rules for the use of personal pronouns in discourse, to which speakers must conform on the pain of unintelligibility. If someone says 'I'm sure he'll be there', then he is referring to a man, and there are anaphoric and cataphoric conventions that partly determine, in the given context, to whom he is referring. But there are no conventions or rules of reference that determine of whom you were thinking when you said to yourself, 'I'm sure he'll be there'.

Paul: Yes, but we *can* say what we think.

Socrates: Of course. You can say that you think that Crito will be at Agathon's mansion tomorrow at sunset for our symposium, but not by reading this off the words you said to yourself. Moreover, if thinking were talking to yourself in the imagination, then if what you said to yourself was 'I hope he'll be there', then in reporting what you think, you would *not* be able to say whom you meant, or when and where you meant.

Alan: Why not?

Socrates: Because, my dear boy, if your thinking *is* your saying whatever you say to yourself, and all you said to yourself is that you hope that he will be there, then that is all you thought. In overt speech, you *can* elaborate whom and what you meant if it isn't clear – given that you were thinking of Crito, of Agathon's mansion and of tomorrow evening. But *if* thinking is no more than discourse with oneself, then it *makes no sense* to ask whom, what and when you meant. *What you thought* is precisely what you said to yourself – no more and no less. As my friend Wittgenstein puts it, thought is the last interpretation.

Locke: I grant you that. But if we think in Ideas and not in Words, as I am inclined to suppose Man to do, then you would perceive what you are thinking, and you could read it off from what you perceive in your Mind.

Socrates: How so?

Locke: You would perceive there the Idea of Crito, conjoined with the complex Idea of walking to Plato's Mansion. The first Idea indicates of what you are thinking, the second what you are thinking of it.

Socrates: But, Master Locke, such a parade of ideas or mental images in my mind could not show me that I expect Crito to walk to Agathon's mansion and not right past it. Might not my mental image of Crito be no different from my mental image of his identical twin? What in this train of ideas would indicate the time at which I expect him? And what shows that we are meeting for a symposium? No, no. Ideas – mental images – can *illustrate* a thought, just as a picture can illustrate a story. But a picture cannot, without a great deal of assistance, *tell* a story – and neither can an idea in the mind. Bethink you: a picture of a hoplite might be a portrait of Aeschylus at Marathon, or it might be just a picture of a hoplite, or it might be a picture of how a hoplite should stand in line, or [*he laughs*] it might be a picture of how a hoplite should *not* stand in line. Moreover, the very same ideas might be present when you *wonder* whether Crito will arrive on time, or when, knowing Crito, you *doubt* whether he will be on time. And these are not *thinking* that he will be on time.

Locke: Ah, … Ah, yes. 'Tis plain 'tis so. I must needs somewhat to alter the Thoughts I formerly had.

Socrates: Good! Excellent! [*He pours himself another glass of wine and quaffs it*] Very good! But we should press on. [*The others pass the bottle of wine around*] Thinking is evidently not a train of ideas passing through the mind. But it is not an inner monologue either.

Paul: Yes. I see. So neither words spoken in the imagination, nor ideas that occur *while* thinking are either necessary or sufficient for thinking.

Socrates: That should be evident, Paul. Look, say something with thought, after reflection, with considered judgement.

Paul: For example?

Socrates: As you please.

Paul: Oh, … All right: What about: 'Alcibiades should not have left Antiochus in charge of the fleet at Notium'.

Socrates: Alas, yes. [*His face clouds over*] Indeed … Still … No matter. Now, do what you did when you said it with thought, only without saying anything!

Paul: Ah, yes. I see … So what *is* saying something with thought, if it isn't accompanying one's speech with anything?

Socrates: Is it not, at least in many cases, simply speech that *takes the relevant factors of the subject into account. Then* it is thoughtful speech.

Paul: Yes … But we often think before we speak, and what is that?

Socrates: It may be no more than pausing to take in what was said by another, and then giving a pertinent response.

Paul: But, Socrates, what makes it *pertinent* is that you have thought up the response – and what is this *thinking-it-up*?

Socrates: Well, suppose Frank says something. You pause for thought, and then respond with a counter assertion that is pertinent. Frank asks you why you think thus, and now you respond by giving reasons for your assertion. Does it matter what happened in your mind? For all Frank cares, Apollo and his chariot may have driven through it. What he wants is for you to *explain* your thought. So what matters is not whether your mind boiled or froze, but what reasons you have for saying what you thoughtfully said. For it is your reasons that make your response to Frank thoughtful.

Locke: Methinks I apprehend the Argument. To give reasons for your Assertion is not to describe a precedent Process *in foro interno*, but to cite a Justification. 'Tis this that is the Cogitation. Or rather, such Cogitation is the *active Power* to do so, whether or not it be exercised.

Paul: I'm starting to feel as if thinking is disappearing into thin air! What about the kind of thinking that consists of reasoning one's way through an argument? And let's suppose that this inner reasoning takes place when you are alone – so nothing at all is said.

Socrates: Need it be more than apprehending – being able and willing to affirm – that such-and-such a conclusion follows from such-and-such premises.

Paul: But surely inferring, reasoning, is a mental process of some kind – a *movement of thought*.

Socrates: Surely! [*Chuckles*] Surely not! A transition in thought is not a movement – how would thoughts move, haha! Where would they move to? Apprehension of a *warrant* is not a kind of movement.

Alan: Socrates, you're losing me. What d'you mean by 'apprehension of a warrant'?

Socrates: To draw an inference in rational *discourse* is to transform a set of enunciated propositions – the premises – into an enunciated conclusion, *in accordance with rules*. To reason one's way to a conclusion *in foro interno*, as Master Locke puts it, is to *apprehend such transformability*.

Alan: But what is this 'apprehending'?

Socrates: Well, my dear boy, it is certainly not an activity of any kind, is it?

Frank: But we know for sure that thinking cannot take place without neural activity in the brain. And *that's* what thinking really is.

Alan: But that's wrong.

Frank: What's wrong about it?

Alan: Well, for one thing, the fact tha' we wouldna be able to think without the right brain activity doesn't show tha' the brain activity is the thinking, any more than the fact tha' we wouldna be able to walk without activity in the motor-cortex shows tha' the brain activity is the walking.

Paul: Good. I agree with that.

Alan: And for another, if thinking – that is: reasoning – is apprehending the transformability of premises into a conclusion, tha's no *activity*. So whatever *brain activity* may be going on while one is thinking, it can't *be* the thinking since the thinking isna an activity at all.

Paul: But Socrates, that still leaves 'apprehending' in the air. What does this apprehending consist in?

Socrates: Why should it consist in anything? But it is clear enough what the criteria are for someone's having apprehended that a conclusion follows from a set of premises. After all, if he asserts the conclusion, and you ask him why – he'll give you the *becauses*! Haha!

Paul: You mean that he will justify his conclusion by reference to the premises and the appropriate rules of inference.

Socrates: Exactly. Good …

[*He takes another drink from his glass and replenishes it. The bottle passes around to the others*] …

Now, let's turn to some other interesting contrasts between activities and thinking. We can measure the time of speaking: when the speaker began to talk, how long he continued to talk and when he stopped talking. But now reflect upon the relation of thinking to time.

Frank: Well, both talking and thinking go on in time.

Socrates: Of course. We've already granted you that, Frank. But now reflect. When you speak with thought, when do you finish thinking the thought you express by your utterance?

Frank: Well, I guess you finish thinking when you finish expressing the thought you think.

Socrates: [*incredulously*] Really? Come now!

[*A brief silence*]

Paul: Oh, I see what you mean. If you hadn't finished thinking before you finished speaking, you would not know what you're going to say until you'd said it! But that is obviously wrong, since if I didn't say what I was thinking, or if I was interrupted while I was speaking, I *do* usually know, and can tell someone, what I was thinking and what I was going to say.

Locke: The want of these sundry Distinctions is the cause of no small Obscurity and Confusion in Men's Thoughts and Discourses. 'Tis plain that things are not as I believed them to be. But it seems very strange in this Doctrine that we cannot say when a Man hath finished his Thought, or even when he has begun it. Pray explain this a little more.

Socrates: Let me put it this way: You can have half a loaf in the pantry, but you can't have half a thought in your mind. [*He chuckles*] We can utter half a sentence, but we cannot think half a thought.

Paul: Of course: 'At sunrise the cat was' is half of the sentence 'At sunrise the cat was hunting mice in the garden', but it is not the expression of half of the thought I express when I utter the whole sentence.

Socrates: Of course not. There is no such thing as half a thought.

Frank: OK. But we can measure how long it took you to think your way to the conclusion of a piece of reasoning. Thinking the thought that is the conclusion of a piece of reasoning may take no time, but the reasoning takes time, and we can often measure how long it took you to come up with the conclusion you reached.

Socrates: In so far as you can *sometimes* specify when I began thinking about a problem, and when I reached a conclusion, then I agree. And in so far as I can be interrupted in the middle of thinking my way through the problem, I also agree. But only thus far.

[*A brief silence*]

Paul: Ah, I see what you are driving at. We can ask someone to solve a complex empirical or practical problem for us, and they may take a few minutes or half an hour to reach their conclusion, on the basis of the available evidence. But a description of what went through their minds while they were thinking their way through the problem is not akin to a description of an ordered sequential activity. And at any given moment, there may be nothing going through their mind – but that does not mean that they have stopped thinking for a moment.

Socrates: Just so.

Paul: Still, the cognitive scientists will say that the information-processing activity and the computational activity occurs in, and is the activity of, the *unconscious* mind. And when it comes to speech and comprehension, that is the doctrine advanced by the most famous linguistic theorist of the twentieth-century, Noam Chomsky.

Socrates: It was advanced long before then by my friend Wittgenstein, when he was writing his first masterpiece, the *Tractatus*. And long before Chomsky advanced this misbegotten idea, Wittgenstein realized that it was completely mistaken. [*He laughs*]

Paul: Why did he reject it?

Socrates: For many different reasons. It is a long a complicated story, and we shouldn't get waylaid by what is no more than a digression into mythologies of the mind and of the mental. But to put matters in a nutshell, it is no more the mind that thinks than it is the brain. It is not my mind that reasons, or that changes its mind – it hasn't got a mind [*he laughs*]. It is not my mind that sees and hears, and it is obviously not my mind that decides that I shall go to the Ambrosian for dinner, any more than it is my mind that has intentions and purposes, that wants and wishes – it is I, Socrates.

Alan: So Thomas Reid was wrong to say that the mind is that in man which thinks, imagines, reasons, wills.[17]

Socrates: Quite so. The mind is not an active agent, even though we often speak of it as a patient.

Paul: What do you mean, Socrates?

Socrates: Well, do we not say: My mind is in a whirl, or in a turmoil, or is wandering or has gone blank? But we don't say: my mind thinks, or my mind reasons or my mind has decided.

Alan: So if it makes no sense to ascribe such agential attributes to the conscious mind, it certainly makes no sense to ascribe them to something called 'the unconscious mind'.

Socrates: Whatever that may be! Haha. I may think and reason, reflect and come to conclusions, but these are not things we can ascribe to the mind or to the agency of the mind, since the mind is no agent. In so far as the mind is *an anything*, it is an array of powers of the intellect and will, and powers are not actors.

Alan: Ay. And these are powers of human beings, not o' their brains or o' their minds.

Paul: But, Socrates, these cognitive scientists have very elaborate theories about the unconscious processes of the mind, about information processing, mental lexicons, meaning analysis units, concept-stores, buffers, conversion units and so forth.

Alan: But these are just fictions. There can be no such thing as a lexicon in the mind – it's just a fancy way of saying tha' you can remember what words mean. And you canna store concepts any more than you can store abilities.

Socrates: Paul, don't let the jargon deceive you. It's just a mind-model that's part of a psychological theory.

Frank: Y' mean just like the hydro-dynamical model of electricity is a model in terms of which to describe electrical phenomena.

Socrates: My understanding of electrical theory is not what it should be, even though Michael Faraday tried to explain it to me. What I was trying to get you to see is that the mind-model of your cognitive scientists is part of the *symbolism* of the theory. It is not what the theory describes, but the means whereby the model represents what is described.

Paul: I'm not sure I follow you, Socrates. Can you elaborate?

Socrates: No, I don't think we should go down that road, Paul. It is a problem for psychology, not for philosophy. The conceptual errors committed when the model is taken to be what it is a model of, and when features of the model are projected onto reality, are not our current concern. Let us pick up the threads where we broke off. We noted that the relation of activities to time is interestingly different from the relation of thinking to time. So although thinking, in some of its forms, is akin to an activity in some respects, in other respects it is not at all like an activity.

Alan: But if thinking is like an activity in some respects, and unlike an activity in other respects, what on earth is it. Thinking's becoming mickle more mysterious, no' less mysterious.

Socrates: No, my dear boy, not at all. It is only as long as you look at it with the prototype of an inner activity in mind that it *looks* as if it is becoming more mysterious. Then, as we point out more and more features which make it *unlike* an inner activity, the more mysterious an activity it appears to be. So stop conceiving of it on the model of an inner activity! The relation of thought to time only looks peculiar if we expect it to be like the relation of activities to time, only hidden from view. But that expectation is unwarranted: [*he chuckles*] Who was it who said that everything is what it is and not another thing?

Alan: D'you mean that there are no inner activities at all?

Socrates: No, no, my dear fellow. You might describe counting goats in order to go to sleep or reciting a poem in your imagination an 'inner activity'.

Paul: Yes ... I see ... So we can conclude that the temporal micro-structure of thinking, so to speak, is altogether different from the temporal microstructure of activities.[18] But that is no mystery, since all it amounts to is that the grammar of thinking is unlike the grammar of activity-verbs in many crucial respects. The distinctions we draw with respect to thinking are very unlike the distinctions we draw with respect to talking, no matter whether aloud or in the imagination.

Socrates: Yes, and that is hardly surprising, since the point and purpose of the concept of thinking in all its multiplicity is completely different from the point and purpose of standard activity concepts. It is bound up with judgement and the reasons for a judgement, with warranted transformations of judgements in reasoning, with qualifications on judgements and with the evaluation of judgements.

Alan: Ay, and with problem-solving and deriving conclusions.

Paul: And doing things with thought is not doing two things at the same time: thinking and acting, but doing one thing thoughtfully, like singing with expression, and not like singing to accompaniment.

Alan: But for all tha', doing something thoughtfully is not a manner of doing it, is it?

Paul: Hmm ... I think one must say: yes, and no! In so far as it is engaging in an activity with intense concentration, then one might say that it is. But in so far as it is engaging in an activity with an awareness of what may go wrong and a readiness to respond to what may go wrong, then doing something thoughtfully moves away from the adverbial prototype.

Alan: It occurs to me tha' these considerations concerning the relation of thought to time surely show tha' thinking isna identical with *any* neural state or process. For neural states and processes are continuous, but thinking is not.

Frank: Well, William James certainly held thinking to be continuous. That's why he coined the term 'the stream of consciousness'. He claimed that consciousness is not, as he put it, 'jointed'. Rather, it *flows*.[19] And that was what the stream of consciousness writers, like Joyce or Virginia Woolf, held too.

Paul: But they conflated what passes through one's mind *while one is thinking* with *what it is that one is thinking*. That's why they wrote ungrammatical, truncated sentences without punctuation to represent their character's thinking, as in Molly Bloom's stream of thought at the end of *Ulysses*. But that, attractive conceit as it may be, *is* confused. For it is, at best, an imagined account of what was passing through Molly Bloom's mind while she was lying in bed thinking. But it is not a super-accurate record of her thinking or of what she thought. Look, if you want to know what Kant thought, you don't want to know what, if anything, was passing through his mind while he wrote *The Critique of Pure Reason*. You want to read *the book*. *That* tells you what Kant thought.

Locke: I would that you explain this a little more particularly.

Paul: Well, we have seen that when one speaks with thought, as we are all doing in this discussion, or when we write with thought and intense concentration, then no mental events need accompany our cogent speech. After all, Mr Locke, if I sincerely tell you that when I am speaking with thought, as I am now, *no* mental images are crossing my mind at all, you will surely not say that I am speaking without thought.

Locke: Why should I not?

Paul: Precisely because, as Socrates just made clear, to speak with thought is, among other things, to speak having taken all the relevant factors into account.

Locke: Ah, yes. I had forgot.

Paul: Now, perhaps an idea or mental image sometimes crosses our mind while we are speaking, but it adds nothing to what we thoughtfully say. What we are thinking is what we say when we sincerely say what we are thinking. And when we think without saying anything, what we think is fully expressed by what we *would* say were we to answer the question 'What were you thinking?' truthfully.

Locke: But this is unintelligible. For to speak Words without ideas is to speak no otherwise than Parrots do. For a Parrot may speak Words only because it has learn'd them, and has been accustom'd to those sounds. But so far as Words are of Use and Signification, so far is there a constant connexion between Sound and the *Idea*; and a Designation, that the one stand for the other: without which Application of them, they are nothing but so much insignificant Noise.[20]

Socrates: You mean, Master Locke, that it is the ideas that accompany speech that make the difference between the meaningless squawks of a parrot and the speech of a human being.

Locke: Of that I am fully persuaded. For Words are the Instruments whereby Men communicate their Conceptions, and express to one another those Thoughts and Imaginations they have within their own Breasts.

Frank: Yeah, I guess that's right. The jargon's a bit dated, but that's what we scientists think too. I mean it's a well-established fact that words are names of concepts. People suffering from dementia or Alzheimer's are unable to recollect words. But at least in the early stages, it's clear that they possess the relevant concept. What's going wrong is failure to access the word-store in the brain that would deliver the name of the concept that is selected from the concept-store.

Paul: What on earth do you mean?

Frank: Look, psycholinguists and cognitive neuroscientists have shown that language processing involves a conceptual preparation module that feeds into the lexical selection module, which is responsible for selecting the appropriate words from the mental lexicon. Hell, all this is old hat – it was shown by Levelt in the 1990s and further developed by Coltheart.[21] And it's obvious that these functions are distributed in the brain, and that was well established by positron emission tomography and fMRI.

Paul: Oh, come now, Frank. We've already rejected that piece of mythology. If we are right – and I am convinced we are – then talk of dictionaries in the brain is no more than a metaphor for our ability to remember words. There is obviously no such thing as storing an ability in the brain. Our ability to remember words is not *in* the brain, but due *to* the brain. And talk of concept-stores is no more intelligible than talk of storing abilities. Concepts are not storables. Rather, to possess a concept is to be able to use a word or phrase that expresses a concept. And the ability to use a word is not to be found in the brain any more than the horse power of a car is to be found under the bonnet.

Frank: Sure. I never said that the ability is *in* the brain. But it's an ability *of* the brain, just as the horsepower isn't in the engine of an automobile, but it is a power *of the engine* of the automobile.

Paul: I agree with you that the horsepower is the power of the engine, and hence of the car. But the ability to learn and remember what words mean is an ability *of the human being*, not of his brain. Of course, I grant you that it is an ability possessed by the human being *because of* the normal functioning of his brain. On the other hand, it has nothing to do with 'concept-stores' and 'mental lexicons' – they're just fictions, as Alan said.

Frank: So how do *you* explain anomia and lethologica, Paul? Hell, these old folk know the concept alright, but they can't specify its name. That's because the concept is stored in the concept module, and the name of the concept is in the mental lexicon, but the connection between them has been broken as a result of neural degeneration or lesion.

Paul: Frank, you chaps really are barking up the wrong tree. Your so-called explanation is merely a re-description of the phenomena in picturesque but misleading form. Look, suppose an elderly person, in mid-speech, can't produce the word 'horse'. So he says, 'She was riding a … a … Oh damn! what's the name of the animal jockeys ride at races?' That shows that he can't bring to mind the word he wants, but he remembers its explanation. The idea that there is a lexicon and a concept-store, and that they have become disconnected is just an alternative way of describing the phenomenon of failing memory. It is, as Socrates explained, a part of a mental model that belongs to the symbolism. It's a means of representation, not a truth about memory. The phenomena of mnemonic failure is that the speaker can remember the meaning of the word he wants to use, but can't come up with the word. That does not show that there actually is a meaning store and a word-store! Meanings are no more storables than concepts, but we can and do remember *what words mean*. To remember what a word means is not to remember an object called a 'meaning', let alone a concept, but rather to be able to answer a question, namely: 'What does such-and-such a word mean?'

[*A few moments silence. Paul pours himself another glass of wine and takes a sip*]

Socrates: Well, that was a most interesting exchange. Very good indeed! Er, Paul, could you pour me another glass of wine too?

[*Paul does so, and hands the glass to Socrates, who quaffs half of it. Paul then fills up the glasses of the others, while Socrates continues talking*]

Now let's get back to Master Locke's worries. Words are not the names of ideas, and they are not the names of concepts either. It is neither ideas nor concepts that animate words, that give them a meaning – that infuse them with life, as my friend Wittgenstein says. What animates words is their use – not what they stand for. What the squawking parrot lacks is not ideas, but mastery of the language. It can't perform any act of speech with the word it squawks, although it may have learnt that a certain squawk in the presence of its owner has a certain effect. But the noises it emits are not symbols, and they mean nothing to the bird. Imagine that a parrot has been taught that when it gives its master a drachma from a pile of coins, it gets a peanut. Has it bought a peanut? Does it know what a drachma is? Does it know its value? Does it know that it is equivalent to four obols? Haha!

Alan: I see. What lies behind a word, and animates it, is not anything in the mind, but its use in the practice of human discourse. And we don't explain what a word means by referring you to an idea in the mind, but by explaining what it means in other words.

Paul: That's very good.

Locke: 'Tis clear that we had puzzled ourselves, without coming any nearer a Resolution of our Perplexities, because we took the wrong course from the beginning.

Socrates: [*raising his finger in the air*] Precisely. [*He chortles*] Was that not what I told you when we began our discussion? [*He laughs*] Now, let's see where we are. We have established, have we not, that thinking is not strictly speaking an *activity* of the mind, nor is it a *process* going on in one's mind while one thinks. Furthermore, we have distinguished between different forms of thinking: the thinking that *informs* physical activities, the thinking that informs thoughtful speech, the thinking that may precede thoughtful speech, the thinking that consists in reasoning and inferring. There are, of course, other forms that thinking may take, but we need not concern ourselves with them now.

[*He pauses to drink his wine*]

Now, the other thing you all seemed to agree on was that thinking is something 'inner'? Why is this such an appealing idea?

Alan: Surely that's obvious, Socrates. We can think something and not say what we think. So my thinking, as well as what I think, is concealed from others. Thoughts are in our mind, and no one else can see into our minds.

Frank: Well, like I said before, I think that thoughts are in the brain, and that's why they're hidden from view. It's the brain that thinks, and if you don't tell others what you think, they can't know what you think, because they can't see into your brain. Or at any rate, they can't see into another's brain without a fMRI scanner. When we've developed these scanners further, and have a good cognitive theory in place, then we'll be able to read off what someone is thinking from the neural activity of their brain.

Locke: I know not how to reconcile these different Opinions. But 'tis clear that we perceive our own Thoughts by Inner Sense or Introspection. Each Man is privileged in having Access to his own Thoughts. For no one can enter the Mind of another to perceive there what he thinks. And is that not why we speak of Thoughts as Inner? If I tell not what I think, no Man can know it. And if I honestly and sincerely say what I think, no Man can gainsay it. I know what I think, and if I tell them not, others must perforce guess it.

Socrates: But Master Locke, you have conceded that if a man were, so to speak, to enter the mind of another who is thinking, he would *not* be able to perceive there what he is thinking. He might, so to say, see a jumble of fleeting images; he might, in a manner of speaking, hear a few exclamations such as 'No, never!' or 'It's her' or 'Yes, I must go there'. But even if he, as it were, heard what he was saying to himself, he would not know whether it was meant or what was meant by it.

Alan: Tha's very good, Socrates. But y' haven't explain'd to us how it is tha' we do actually ken what we're thinking. I mean, we all agreed tha' we know what we think by introspection! Mr Locke called it 'perceiving our perceptions' or 'inner sense'; Leibniz called it 'apperception' and William James called it 'introspection'. But you've argued and almost persuaded us tha' no matter what we introspect, it isn't the real thinking.

Paul: Yes, I agree with Alan, Socrates. Inner sense or introspection was supposed to explain just that. You've taken us out of the frying pan, only to drop us into the fire. How, on your account, do we know that we're thinking or what it is that we're thinking?

Socrates: [*laughs*] Yes, I know that's what it looks like. We have two problems on the table now. First, is it correct to say that thinking is, in some sense, 'inner'. And secondly, how do we know that we are thinking and what it is that we think. Hmm. [*He pauses for thought*] Perhaps we should handle the second question first.

Locke: Surely, common Experience makes it Impudence to deny that it is by inner Sense that we know what we think.

Socrates: Tell me, Alan, how did Professor James define 'introspection'?

Alan: Ay, I remember tha' clearly. I heard him give tha' lecture at the Athenaeum Lecture Theatre here last year. He said that introspection means 'the looking into our own minds and reporting wha' we there discover'. And he remark'd that wha' we discover, when we introspect, are states of consciousness. I remember he added a purple passage. Tha' we have *cogitations* o' some sort, he said, is the ultimate unshakable certainty in a world whose other facts have tottered in the face o' philosophical doubt. *All* people, he said, unhesitatingly believe tha' they *feel* themselves thinking, and tha' they distinguish the mental state as an inward activity or passion. He added tha' he considered this belief as the most fundamental of all the postulates o' Psychology.[22]

Locke: Most curious. Master James must have learnt much from Monsieur Des Cartes.

Socrates: Excellent. Now we have something into which we can sink our teeth [*he chuckles with glee*]. Now do you really *feel* yourself thinking? What with? With your finger tips? Hmm! And where do you feel the thinking? In your head? Is this feeling like a headache? Or is it like an itch? Can you scratch it?

Locke: No, no. We perceive the Operation of the Mind that we denominate Thinking, and so too, we perceive, at the same time, the Object of Thinking.

Socrates: And pray what do you perceive it with? With the eyes of the mind? So you *see* your thoughts – and don't feel them? And if it is with the eyes of the mind, is the light in there always brilliant sunshine, eh? And how could you *see* a thought? Can you also *see* *thinking*? I know what it is to see *someone* thinking. He may perhaps frown, and put his chin on his fist and concentrate. But how, Master Locke, would you see thinking in the mind, eh?

Locke: No, no. If it shall be demanded with what Sense Organ I apprehend the Operations of my Mind, then I think the true Answer is that 'tis not with a Sense Organ that I apprehend my own Thoughts.

Socrates: But you do 'apprehend' them?

Locke: 'Tis plain 'tis so – otherwise how could I say what I think?

Socrates: Ah, now we reach the nub of the matter. Is it not enough that you think things to be so, and then you say what you think?

Locke: But I must certainly know what I think before I say what I think? I would be glad to learn from you how I could say what I think if I did not know it.

Socrates: Take it slowly, Master Locke. You are now in very treacherous seas, and if you don't steer your galley carefully, you'll end up on the rocks. Reflect a moment. When *do* you say that you *don't* know what you think?

Locke: [*pauses*] … I think the Answer must be: when I am undecided what to think.

Socrates: Precisely. So it is not like not knowing what another thinks. I mean, if you see our friend Plato sitting at a table with his quill paused, with a far away look on his face, and his brow furrowed, then you may know *that* he is thinking something over. But you don't know *what* he thinks. But when you tell me that you don't know what *you* think on some topic, it is *not* a matter of your definitely thinking something, but not knowing what it is that you think.

Locke: No … 'tis strange. For no one would pronounce that he doth not know *that* he thinks things to be so. He would rather say that he knows not *what* he thinks.

Socrates: So if someone, confronting a problem, does not know what he thinks, that does not mean that he thinks something but knows not what?

Locke: No, no. That would be mere Confusion. If a Man knows not what he thinks, that signifies that he knows not WHAT TO think.

Socrates: Quite so. And if he knows not what to think, does he then *introspect*, or *perceive his perceptions* in order to discover what he thinks?

Locke: Ah, I see where you are leading me. No, when a Man knows not what he thinks concerning a disputed Question, he must turn to the Evidence that speaks in favor of one Opinion and compare it to the Evidence that supports the other. Then he must decide.

Socrates: Exactly. And when he decides, does he not say 'Now I know what I think'?

Locke: Ah, yes. So 'I know what I think' signifies that I have come to a conclusion, have decided what TO think. It does not signify that I have perceived my Thoughts and can say what they are. It signifies that I have deliberated and come to a Conclusion, and THAT is why I can say what I think.

Paul: Socrates, do you then mean that we *don't know* what we think when we think something to be so?

Socrates: No, my dear fellow. You don't know what you think when you are undecided – when you don't know what to think. When you say 'Now I know what I think', that is an exclamation that you have decided what to think. If you say to someone that you know what you think but you are not going to tell him what you think, what that means is that you can say what you think but you are not going to.

Alan: So you jus' have to think thro' whatever problem concerns you, and then say wha' you've concluded. So perceiving your own thoughts does na come into it.

Frank: I still have to wrap my mind around this. Do you mean that there's no such thing as introspection?

Socrates: No, of course I don't. But what I suggest is that the way you represent introspection is mistaken. Reflect that when a potter *throws a pot* he does not throw anything at anyone [*he laughs gleefully*], and introspection, despite Professor James, is not a matter of looking into one's mind. [*He chuckles*] It is a matter of reflection on one's character and behaviour, and on the reasons for one's passions and attitudes.

Paul: Yes, *of course*. It is the Marcel Prousts of the world who are introspective [*he laughs*] – and their introspective endeavours are none too reliable either! Being able to say that one thinks this or that involves no introspection. It involves only coming to a conclusion, that is: thinking, making a judgement, and saying what one has concluded.

Locke: I am persuaded. I confess that I confus'd the ability *to say* for the ability *to see*. To say what one thinks, one need not perceive one's thinking. One need but think and say what one thinks.

Paul: No, Mr Locke. It is not that you *need not* perceive your perception. The only perception you *can* perceive is another person's perceiving. There is no such thing as perceiving your own perception.

Locke: This I must concede. I stand corrected.

Socrates: Good. Now let us not lose sight of our quarry. We have not yet got to the bottom of the idea that thinking is something *inner*, which only the thinker can see, and others can merely surmise.

Locke: But are a Man's Thoughts not hidden within his Breast?

Socrates: Is that what you think, Master Locke?

Locke: Yes, it is.

Socrates: But there is nothing hidden about it – you just told us what you think, and now we all know.

Locke: Yet had I not told you, you would have been ignorant.

Socrates: But you did tell us. So nothing is hidden!

Locke: But had I not told you, I would have known and you would not.

Socrates: I thought we just agreed that had you not told us, you *could* have told us – but not because *you knew what you thought by introspection*, but because you can *say* what you think. You can *express* the conclusion you come to after thinking – namely: that things are thus-and-so. And, of course, you can tack on an 'I think' in front of it. But there is here no room for knowledge or ignorance in the manner in which there is in the case of the thoughts of another.

Alan: But surely, Socrates, we do sometimes hide our thoughts from others.

Socrates: Yes. You might keep your commonplace book or diary under lock and key. That would be hiding your thoughts.

Locke: 'Tis true. And a Man might write his Thoughts in code.

Socrates: Exactly. But merely not saying what you're thinking is no more hiding anything than not saying what you're currently perceiving is hiding anything.

Paul: Yes. That's very good. We confuse not-saying with concealing, and not being expressed with being hidden. All that is true is that others *may* not know what I am thinking if I don't tell them.

Socrates: Well, a little more than that, my dear friend. If you don't tell another what you are thinking *in order* that they should not know, or in a context in which they would naturally expect you to tell them, then too you might be said to be concealing your thoughts.

Paul: I see. You're right, of course. But it is still worth noting that if our thoughts were 'metaphysically hidden', as it were, then it would not be so difficult to conceal them – as it often is. Nor would it often be so easy to see what someone is thinking if you know them well and watch their face. [*He laughs*] I often know what my small son Thomas is thinking before he does!

Alan: But for a' that, Paul, don't we have privileged access to our own thoughts? After all, I *do* often have to *ask* you what you're thinking about. I mean, we say 'A penny for your thoughts'.

Locke: If I have correctly apprehended Socrates, then Access is here a misapply'd Word. A Man may have access to a lock'd Chest if he have the Key. He may have Access to a Library if he be permitted to enter it and to use it. He may have Access to the King if he be granted the Privilege to see him. But, if Socrates be right, no Man hath *Access* to his own Mind. For we are deceived by the Use of this Word. To be able to think and to say what one thinks or no as one pleases is not having Access to aught.

Socrates: Precisely. [*He beams and rubs his hands, and then pours himself another glass of wine*] Do we not confuse knowledge derived from privileged access – as when a man peeps into a closed box that others cannot look into – with having a privileged word?

Alan: I canna follow you, Socrates.

Socrates: Well, there is nothing incredible and nothing that calls for a special explanation if a man's expression of his own thoughts has a privileged *status*. After all, the thoughts that he honestly expresses are *his* thoughts. For remember, my sincerity does not guarantee the truth of my description of *your* thoughts, but when I tell you sincerely what *I* think, then *truthfulness guarantees truth*. I have no *access* to my thoughts,

let alone privileged access to them. My thoughts aren't *hidden* somewhere, where only I can look. They are *made*, not found. *My thoughts
are the thoughts I think.* And if I truthfully say what I think, my word,
normally, overrides the words of others regarding what I think. And
that is not because I can peep into a closed box into which others
cannot look.

Frank: OK. Maybe you're right that just not saying what you're
thinking isn't hiding anything. But still, the fact is that thoughts are in
your brain. Now I agree that the brain is not a 'metaphysical' hiding
place. But you can't see what's in the brain unless you use a scanner.

Paul: Frank, you're obsessed with your fMRI scanners.

Frank: Well, if I am it's because of what they teach us. It's the brain
that thinks, and we can see it thinking when we make use of an fMRI
scanner, and we can actually see where it's thinking.

Socrates: But Frank, is it the brain that walks? Surely you walk with
your legs, not with your brain?

Frank: Yeah, OK. But it's the brain that thinks.

Socrates: Hmm. Well, it is true that thought needs a subject – a
thinking thing.

Locke: 'Tis true. It is upon this that the esteemed Des Cartes insisted.
Yet whether this Substance be Matter or Spirit is a Source of many
Perplexities.

Socrates: And how do we identify a thinking subject?

Alan: Ay, ... I see where you're going. We dinna say that a table thinks,
or that it is thoughtless. And we dinna say that a heart thinks, or that it
dinna think. Hearts pump blood – they're not in the thinking business.
It dinna make sense to say that a table, or a heart, is thinking. So
you're suggesting that it dinna make sense to say tha' the brain is
thinking either?

Paul: Yes, that's dead right. We only say that something thinks if we
know what it *would be* for it to display thought in what it says and does.
For a being to be a thinking creature, its normal behavioural repertoire must include thoughtful behaviour and behaviour that expresses
thought. Then it also makes sense to say that it *hasn't* thought or *isn't*
thinking. So we don't say that a table *isn't* thinking, any more than we
say that it *is* thinking. For tables can't do anything that would *count* as
manifesting thought. Indeed, tables don't *behave* at all. So they neither
think nor fail to think – it *makes no sense* to ascribe thought or lack of
thought to them.

Locke: So 'twould be an *Abuse of Words* to say that the Brain thinks or
reasons. Such an Application of these Words must needs cause great

Disorder in Discourses and Reflections about the Brain and Thinking, and be a great inconvenience in our Communication by Words.

Frank: OK. I can vaguely see what y'all are driving at. But you can't deny that we think *with* our brains, and that thinking takes place *in* the brain. Hell, y'can *see* it.

Socrates: Don't be so hasty, my friend. You see with your eyes, do you not?

Frank: Yeah, sure.

Socrates: And you hear with your ears. You must grant me that.

Frank: Sure.

Socrates: And what do you *do* with your eyes in order to see with them?

Frank: Well, [*he laughs*] I guess I'd better open 'em.

Socrates: And to see better?

Frank: I'd go closer to look better, if that was possible. Or turn on the light to increase visibility. Or put my eye to the eye-piece lens on a microscope or telescope.

Socrates: Yes. And now what do you do with your ears? I mean: to hear better with them.

Frank: Oh, I see. And now you're gonna ask me what I do with my brain in order to think, or to think better.

Paul: Yes, of course. Because the brain is not an organ of thinking in the sense in which the eyes and ears are severally organs for seeing and hearing. You can't *do* anything with your brain in the sense in which you can do something with sense organs. And although you can think better when it's quiet than when it's noisy, there are no thinking conditions in the sense in which there are visibility conditions or auditory conditions.

Frank: So y'mean we don't think with our brains?

Paul: No, not in the sense in which you see with your eyes and hear with your ears. Only in the sense in which you walk with your brain.

Frank: What d'you mean?

Paul: I mean that just as you need a brain in order to walk, but you don't walk with the brain, so too you need a brain in order to think, but you don't think with your brain. You don't think *with anything*.

Frank: But surely you think *in* your brain.

Socrates: No, my dear fellow. What you see in this machine of yours is whatever takes place in the brain when you're thinking.

Frank: So where the hell *is* the thinking?

Paul: Look, Frank. What you see on a scanner is an increase in the blood oxygen level dependency signal in the frontal cortices that takes place while the subject is thinking. But the place where thinking goes on is not in the brain, but wherever the thinker is when he is thinking. You've been perceiving – that is: listening to – thinking for the last hour. And you've participated in it, to boot. If the phrase 'seeing thinking' means anything then that is what it means. If you want to see thought, then listen to and participate in a really good conversation … Like the one we've been having, thanks to Socrates.

Socrates [*chuckles, and leans forward to finish his wine*] Good. Yes, that *was* a good conversation. I positively enjoyed it. Now [*he gets up*] let's all go to dinner at the Ambrosian.

Alan: But Socrates, we have na discussed the relationship between thought and language. And we have na spoken about animal thought either.

Socrates: [*starting to move away*] Dinner time, my dear fellow, dinner time. Haha. We can carry on later.

Frank: [*as he rises to his feet, he addresses Paul*] Hell, I'm still not sure. I mean, give us a few years more, and we'll be able to read off what people are thinking from the scanner. So how can the old boy say that thoughts aren't in the brain?

Paul: [*patting Frank on the shoulder*] Well, you'll just have to chew on it until after dinner, old chap!

[*They all walk off into the trees*]

Notes

1   Susan Greenfield, in one of her television broadcasts, pointing at a brain image on a fMRI scanner, remarked, 'Here, for the first time, we can actually *see* thinking.'
2   Leibniz, *Principles of Nature and Grace*, §4.
3   Sir Matthew Hale, *The Primitive Origination of Mankind* (1677).
4   C. Blakemore, 'Why does the inner eye see so little? It gives us only a tiny glimpse, and distorted at that, of the world within. Much of what our brains do is entirely hidden from consciousness … Most of the actions of the human mind are beyond the gaze of the inner eye of consciousness.' *The Mind Machine* (BBC Books, London, 1988), p. 14; C. Koch, 'Much of what goes on in the brain bypasses consciousness', in *The Quest for Consciousness* (Roberts, Englewood, Colorado, 2004), p. 3. See also F. Crick, *The Astonishing Hypothesis* (Touchstone, London, 1995), p. 266 and *passim*.
5   A view briefly adopted by Wittgenstein in the early 1930s but later abandoned.
6   Plato, *Theaetetus* 189e.
7   Plato, *Sophist*, 263e.
8   Sophocles, *Ajax*, ll. 365–68.
9   Locke, *An Essay Concerning Human Understanding*, III-i-2, III-ii-1
10   Hobbes, *Leviathan*, chap. iv.
11   Hobbes, *Human Nature*, chap. 5, §14.

12   Arnauld, *The Art of Thinking*, Part II, chap. i.

13   Locke, *Essay*, III-ii-1 (abbreviated)

14   P. T. Geach, 'What Do We Think With?' in *God and the Soul* (Routledge and Kegan Paul, London, 1969), p. 31.

15   Severin Schroeder, 'Is Thinking a Kind of Speaking?', *Philosophical Investigations* 18 (1995), p. 146.

16   E.g., P. N. Johnson-Laird, *Mental Models* (Harvard University Press, Cambridge, MA, 1983); R. Jackendorff, *Consciousness and the Computational Mind* (MIT Press, Cambridge, MA, 1987).

17   Thomas Reid, *Essays on the Intellectual Powers* [1785] (Edinburgh University Press, Edinburgh, 2002), p. 20.

18   Schroeder, 'Is Thinking a Kind of Speaking', p. 148.

19   Italics in original. W. James, *The Principles of Psychology* (Holt, New York, 1890), vol. I, p. 239: 'Consciousness, then, does not appear to itself chopped up in bits. Words such as "train" or "chain" do not describe it fitly as it presents itself in the first instance. It is nothing jointed; it flows. A "river" or a "stream" are the metaphors by which it is most naturally described. *In talking of it hereafter, let us call it the stream of thought, of consciousness, or of subjective life.*' Thomas Reid anticipated James. He wrote: 'the stream of thought flows like a river, without stopping a moment', *Essays on the Intellectual Powers of Man* [1775] (Edinburgh University Press, Edinburgh, 2002), p. 420.

20   Locke, *Essay*, III-ii-7.

21   W. J. M. Levelt, 'Accessing Words in Speech Production: Stages, Processes and Representations', *Cognition* 42 (1992), pp. 1–22.

22   James, *Principles of Psychology*, I, p. 185.

SUPPLEMENTARY READING

P. M. S. Hacker, *The Intellectual Powers: A Study of Human Nature* (Wiley-Blackwell, Oxford, 2013), chap. 10.

Gilbert Ryle, *Collected Essays*, vol. II (Hutchinson, London, 1971), essays 19, 22, 30, 34, 36.

Gilbert Ryle, *On Thinking* (Blackwell, Oxford, 1979), essays 1–5.

Severin Schroeder, 'Is Thinking a Kind of Speaking?', *Philosophical Investigations* 18 (1995), pp. 139–50.

Wittgenstein, *Philosophical Investigations* [1953], fourth edition (Wiley-Blackwell, Oxford, 2009), §§316–62.

For help with Wittgenstein's difficult text: P. M. S. Hacker, *Wittgenstein: Meaning and Mind*, volume 3 of *An Analytical Commentary on the Philosophical Investigations*, extensively revised second edition (Wiley-Blackwell, Oxford, 2019), Part I, *Essays*, Essays I, III, VI, X, XI; Part II, *Exegesis* §§316–62.

Seventh Dialogue

THOUGHT AND LANGUAGE

Protagonists:

Socrates: gravelly voice, slight regional accent. Dressed in ancient Greek manner. (It should be noted that Socrates has spent a great deal of time talking with his friend Wittgenstein.)

Paul: a middle-aged Oxford don of the 1950s, dressed in well-cut sports jacket, waistcoat and tie, Oxford English accent.

Alan: a Scottish post-doc, tweed jacket and woollen tie, soft Scots accent.

Frank: a contemporary American neuroscientist in his forties, casually dressed, American accent.

John Locke: in seventeenth-century scholar's garb. A pedantic and slightly reedy voice.

The scene is a garden in Elysium in the late evening. The moon is bright. A rich ver-dant lawn is surrounded by flower beds and rose bushes in bloom, with tall trees behind. Beyond, there is a moonlit view of lake and mountains. There are five garden chairs around a low table on which there are some scattered books. There are candles in two large candelabras on the table, and lanterns behind the chairs. The noise of laughter and animated discussion is audible as the participants return from dinner. They take their seats.

Socrates: I must say, they do one proud at the Ambrosian. That was an excellent dinner. Now, where were we?

Paul: Well, you showed us something none of us really expected, Socrates, namely that thinking is not an activity – just like a physical activity, only mental. I must admit that I had not anticipated that.

Alan: Ay. It goes against one's intuitions.

Socrates: [*chuckles*] Since one's intuitions are just one's ill-informed and unreflective hunches and guesses, that shouldn't worry us.

Alan: Ay, but it's damned painful to have one's ideas pulled up by the roots.

Socrates: Well, my boy, the pain is a small price to pay for getting rid of the weeds. [*He laughs*]

Paul: You showed us that the relation of thinking to time is altogether unlike the relation of a physical activity to time – that in an important sense it is misleading to construe thinking as an activity. And you showed us how misleading is the idea that thinking is something 'inner' and private to the thinker, and how misguided the notion of introspection is when construed as apperception – as 'inner sense'.

Frank: Yeah. And then you got the others to say that it isn't the brain that thinks, that we don't think with our brain, or in our brain, for that matter. I have difficulty with that. It goes against the grain of current cognitive neuroscience, y'know.

Locke: Yet to go against the Grain is often the surest Path to Wisdom. For the Doctors of the Schools, aiming at Glory and Esteem, for their great and universal Knowledge – easier a great deal to be pretended to, than really acquired – found it a good Expedient to cover their Ignorance with a curious and unexplicable Web of perplexed Words, and procure to themselves the admiration of others by the Abuse of Terms, the apter to produce Wonder, because they could not be understood.

Frank: OK. I'm not as convinced as the rest of you guys. But let's move on. You've argued that what goes on in the mind while we're thinking is not the thinking. I still don't follow you. We surely think *in* some language or other. Hell, whatever you say, we *do* talk to ourselves in our imagination, and what we say, even if it isn't always what we think, often-times *is*. And when we learn a foreign language, don't we say such things as 'I can speak French, but I can't think in French' or 'My French is now getting better: I can even think in French'?

Paul: Your first worry is easy to resolve, Frank. Of course you can say to yourself what you think, just as you can say out loud what you think. But what makes it what you think is not that you say it to yourself, since you can say lots of things to yourself that you don't think, just as you can say lots of things aloud that you don't think.

Frank: So what makes something I say to myself an expression – an inner expression, I guess – of what I think?

Paul: Exactly what makes what you say out loud an expression of what you think. If what you say to yourself is your firm opinion or unwavering belief, then what makes it what you *opine* or *believe* is your commitment *to it, your drawing inferences from it and acting on them*, and so on. On the other hand, if it is a qualified judgement that you, as it were, express to yourself, then it is your recognition that you are not in a position to exclude all doubts, that your grounds are inconclusive, but sufficient, other things being equal, to provide an adequate reason for thought or action. SO it's what you *think* to be so, not what you know to be so.

Alan: That's fine. But what about Frank's second point. It's surely right that if you dinna speak a foreign language very well, then you'll say that y' canna think in that language. And when your mastery of the language is adequate, you'll say that you now even think in French, or German, or whatever.

Socrates: So far, so good, Alan. But what does it mean? [*Silence for a few moments*] Come now my friends, is it not obvious?

{ **Paul:** Anything but obvious.
{ **Alan:** No, not obvious at all!

Socrates: [*chuckles*] It really is, you know. If someone says 'I can speak Persian, but I can't think in Persian, that means that before he can say anything in Persian, he must first decide what he wants to say (and *be able* to say it in his native tongue), and then to struggle to find the right Persian words to express what he thinks. But it does not follow that one can say *of a native speaker* that he thinks in his mother tongue – unless that just means that when he talks to himself in the imagination, he does so in his mother tongue.

Locke: But how are we to give a right view of the Case when a Man declares that he converses in French so proficiently, that he even thinks in French?

Socrates: Gentlemen, you must think! Haha. All it means is that his mastery of the language is so well advanced that he does *not* have to think what he wants to say, and then pause to try to recollect the foreign words in which to say it. Rather, he can give expression to his thoughts in the foreign language without any hesitation.

Alan: That's good. But what about the case when someone says 'That's exactly what I was thinking, only I couldna find the right words t' say it'? Is that not a case of thinking without words and then finding *from another fellow* what words express one's thought?

Paul: Yes, that is an interesting case, Socrates.

Socrates: My friends, don't let yourselves be so readily deceived by words. They are wonderful tools, but you must grasp them by the handle if you want to use them. In such cases, it is not that the other person finds the right words to translate your wordless thought. It is rather that he finds the right words to match the phenomenon you too were trying to describe but could not readily do so, and he describes it in a way that you find appropriate. *That's* why you say 'That's what

I was thinking' or 'That's what I was trying to say, only I couldn't find the right words'.

Frank: Yeah. I see. OK, I accept that. But that's just idiom. The fact is that everyone agrees that we think *in* something.

Paul: Who's everyone?

Frank: Well, only such guys as Francis Galton, Albert Einstein and Roger Penrose.

Alan: That's very interesting, Frank. So what did they say?

Frank: Well, you can find some of it here, in this book by Hadamard. [*He leans forward and picks up one of the books on the table*] He was a French guy, a mathematician, who went around asking great creative thinkers, especially mathematicians, how they thought, what they thought in and so forth. [*He opens the book at a marked page*] So, when Einstein was asked, he wrote: 'The words of language as they are written or spoken, do not seem to play any role in my mechanism of thought. The psychical entities which seem to serve as elements of thought are certain signs or less clear images which can be "voluntarily" reproduced and combined.'[1]

Paul: Hmm. Interesting. And what about Penrose?

Frank: [*putting down the book and picking up another one*] Well, he wrote about it in his book *The Emperor's New Mind*. [*He flicks through the pages*] Yeah, I got it. This is what he wrote: 'Almost all my mathematical thinking is done visually and in terms of non-verbal concepts, although the thoughts are quite often accompanied by inane and almost useless verbal commentary … the difficulty that … thinkers have had with translating their thoughts into words is something that I frequently experience myself. Often the reason is that there are simply not the words available to express the concepts that are required. In fact, I often calculate using specially designed diagrams which constitute a shorthand for certain types of algebraical expression … This is not to say that I do not sometimes think in words, it is just that I find words almost useless for *mathematical* thinking.'[2] Unquote. Now, that seems to me to be definitive.

Alan: So they say that we have to think *in something* but not in words?

Frank: Well, y'know, Francis Galton went even further. What he wrote was, let me find it [*he pages through the book*] … yeah, here it is. What he wrote is, 'It is a serious drawback to me in writing, and still more in explaining myself, that I do not think as easily in words as otherwise. It often happens that after being hard at work, and having arrived at results that are perfectly clear and satisfactory to myself, when I try to express them in language I feel that I must begin by putting myself upon quite another intellectual plane, I have to translate my thoughts into language that does not run evenly with them. I therefore have to waste a great deal of time in seeking appropriate words and phrases.'[3] Unquote. Right! Now it seems to me that the testimony of these guys makes a pretty good case for the idea that we have to think in something. And there seems to be a fair amount of agreement that we don't think in words but in images or non-linguistic concepts.

Locke: This, methinks, doth confirm what I wrote in my *Essay*.[4] The Thoughts of Man, being composed of Ideas received from Experience, are all within his own Breast, invisible and hidden from others. Nor can they, of themselves, be made to appear. But the Comfort and Advantage of Society is not to be had without Communication of Thoughts. Hence it was necessary, that Man should find out some sensible Signs, whereby those invisible Ideas, which his Thoughts are made up of, might be made known to others. For this purpose, nothing was so fit, either for Plenty or Quickness, as those articulate Sounds, which with so much Ease and Variety, he found himself able to make. Thus Words come to be made use of by Men, as the Signs of their Ideas.

Socrates: So, you seem to agree that we must think in something.

[*He pauses for any sign of dissent. There is none, so he continues*]

So, if you want to exclude 'thinking in words', the candidates for what we are conceived to think in seem to be *ideas*, as Master Locke suggests, or *non-linguistic concepts*.

Frank: Yeah, that's right. And many cognitive neuroscientists agree. Antonio Damasio seems to agree with Mr Locke. Damasio wrote that language is a translation of something else, a conversion from non-linguistic *images* which stand for entities, events, relationships and inferences. In his view, language operates by symbolizing in words and sentences what exists first in non-verbal form.[5] By contrast, Edelman – a Nobel laureate – holds that *concepts* precede language. On his view, concepts are constructs the brain develops by mapping its responses prior to language. Language then develops by epigenetic means to further enhance our conceptual and emotional exchanges.[6] Now that fits what I said before dinner: concepts are stored in the brain in a concept module, and words, which are stored in the mental lexicon that is realized in the brain, are names of concepts. But you guys didn't like that.

Paul: No, we didn't. And for jolly good reasons too.

Socrates: Come now, let's not retrace our steps over ground that we have already covered. When Einstein said that words of language, as they are written or spoken, don't play any role in his mechanism of thinking, does that mean anything other than the fact that, on the whole, when he is thinking hard, he does *not* talk to himself in the imagination?

Alan: No, I dinna see how it could mean anything more than that.

Socrates: And when he said that the psychical entities, as he put it, which seem to serve as elements of thought, are certain signs or less clear images, can that mean anything other than the fact that when he is thinking, some signs and vague images cross his mind?

Alan: No – and we've already argued that that isn't his thinking, and 't isn't what he is thinking either.

Socrates: So when Penrose says that when he is working on mathematical problems, he does not think in words because there are not words to express the concepts he needs, does that mean more than that he does not, when thinking about mathematics, translate complex algebraical formulae into English.

Alan: That's right. Although it seems that he finds it useful, while thinking, to visualize formulae or diagrams.

Socrates: Yes, but that does not mean that he thinks *in* mathematical formulae in the sense in which he speaks in English. It merely means that it is a valuable heuristic to visualize them. But now, he said something much more interesting than such a commonplace. He spoke – correct me if I am wrong, Frank – he spoke of translating thoughts into words. Now, how would one go about *translating* an unexpressed thought into words?

[*A brief silence*]

Paul: Yes, I see what you mean. He is confusing *expressing a thought* with *translating a thought*. One can't *translate* unexpressed thoughts, since there is nothing *to* translate – only an unfulfilled *possibility* of expression.

Frank: But what about Galton? He said that he arrives at results that are perfectly clear, which he then has to translate into language.

Paul: But is that not like knowing what one wants to say before saying it? That often happens to all of us, but that doesn't mean we are translating thoughts into words.

Socrates: Bravo – well said. It is obvious enough that the difficulty to which he alludes is not that of translating, but of expressing.

Frank: But Edelman claimed that we think in concepts, that we form our judgments by means of non-linguistic concepts, and then translate the judgment into words. He claims that concepts precede language. Concepts, according to him, are constructs the brain develops by mapping its responses prior to language. So y'can think in concepts and then translate what you think into words.

Locke: This I cannot accept. But it hath greater Plausibility to contend that the Instruments of Thought are Ideas, which are given in Experience, or constructed by the Mind out of Ideas given in Experience.

Paul: I don't think that it makes sense to speak of pre-linguistic concepts. Concepts are no more than abstractions from uses of words. To possess a concept is to have mastered the use of a word. There is no such thing as the brain possessing, let alone constructing, concepts, since there is no such thing as the brain mastering the uses of words. And heaven only knows how anyone, let alone the brain, might develop concepts by mapping brain responses prior to language. It is human beings, not brains, who master the use of words in the course of their acculturation into a linguistic community. To master the uses of words is not to learn a list of names, least of all names of concepts. Concepts don't *have* names anyway. To acquire concepts, to master the use of words expressing concepts, is to learn how to *do* something, namely to engage in the language-games of a linguistic community.

Alan: But that doesna show that concepts are not independent of, and in some sense prior to, language.

Paul: Surely that Platonist notion is incoherent, Alan. One could say — although it might be a little misleading – that concepts are techniques of using words, or ways of using words. It is surely obvious that it makes no sense to speak of techniques of using words existing independently of words! That would be like saying that the use of a tool exists independently of the tool – as if the use of a screwdriver might exist antecedently to screwdrivers. [*Socrates chuckles*] We find it useful to talk about concepts, over and above talking about the meanings of words, when we wish to abstract from parochial, language-specific features and to speak of distinctions that transcend local linguistic differences. But the suggestion that concepts exist independently of words and phrases and their uses is patently absurd. What on earth is it supposed to mean?

Socrates: And what of ideas, Paul?

Paul: Well, we discussed that before dinner. In so far as ideas are supposed to be sense-impressions, then thinking is not an operation with or on ideas. Thoughts are not composed of ideas, nor indeed of concepts for that matter. They are not composed of anything. To be sure, one must have mastered many concepts, learnt the use of many words, to be able to think anything but the most rudimentary thoughts. And equally, one must have, and have had, a multitude of perceptual experiences or 'ideas' in order for one to be able to think. But ideas are not the medium of thought.

Locke: If Thoughts be not composed of Ideas, and if we do not translate Thoughts into Words, then it is a Matter of great Obscurity whence they derive. If expressing a Thought be not translating a Thought into Words, how then doth a Man think antecedently to expressing his Thought?

Socrates: Master Locke, you must surely admit that if you perceive things to be so, you can say so.

Locke: This I must concede.

Socrates: But if you can *say* how things are, you can also *think* how things are without saying that they are so. You can think a thought, that is, it makes sense to say of someone that he has a certain thought, only if he *can* express it – no matter whether he does so or not.

Frank: I don't get it. What then is the medium of thought?

Socrates: Why should thought *have* a medium?

Frank: I don't follow.

Socrates: Well, perhaps we should approach this one in a roundabout way. We need to reflect more upon the relationship between thought and language.

Frank: OK. Y'can think without speaking.

Alan: [*with a chuckle*] As well as speak without thinking.

Socrates: Of course, my dear boy. But what *is* speaking without thinking?

Frank: Well, if you weren't here, I'd have said that it is producing sentences without any inner process or activity of thinking going on. But you showed us, before dinner, that thinking is not an accompaniment of thoughtful speech. So speaking without thinking can't be speaking without an inner accompaniment.

Paul: No. It's now clear that speaking without thinking can be different things. It may be speaking without taking some relevant factor into account. It may be making various kinds of trivial mistake, such as giving someone directions how to get somewhere from one's own house instead of from theirs – so that what one says is irrelevant. It may be saying something mechanically, such as repeating a message or dictating something from a text. It may be giving a routine talk while thinking of something else. All these, and I'm sure much more, count as speaking thoughtlessly or without thinking. Come to think of it, speaking thoughtlessly and speaking without thinking are not the same.

Frank: Why not?

Paul: Speaking thoughtlessly is commonly speaking without adverting to the circumstances of the utterance or to the person whom one is addressing, and so unintentionally making an offensive or insensitive remark.

Socrates: Good. Now that you have got those niceties out of your system, Paul, let's get back to the issue we need to look at. Does the *ability to think* presuppose mastery of a language, or is thought prior to speech?

Frank: You mean: does one have to be able speak before one can think? So that babies and animals can't think? Well, I've already told you what most neuroscientists say. For the most part they think that language presupposes thought.

Locke: That is a Notion like to those of the Master-philosophers of my Times. 'Twas not only I who advanced the Idea that Thought be prior to Speech. I have already recounted to you the opinion of Master Hobbes, as well as that of Monsieur Arnauld. They held that the general use of speech is to transfer our mental discourse into verbal.

Paul: Indeed, it is most interesting that something akin to this conception is embraced not only by linguistic idealists of your persuasion, Mr Locke, but also by your fiercest critic in the nineteenth and early twentieth centuries, Gottlob Frege.

Locke: Ah, I have heard of this Thuringian gentleman. A logician, methinks? From Jena?

Paul: Yes, quite so. He held that if we attend to the true nature of thinking, we shall not derive thinking from speaking, for thinking will emerge as that which has priority.[7] Thoughts, in his view, are abstract objects that exist independently of us and of our minds. Thinking is not creating thoughts, but *grasping* them. Grasping he thought to be a special and mysterious mental process which connects the psychological with the logical, that is to say, the psychological process of thinking with the logical objects constituted by thoughts, conceived as abstract, sempiternal objects.[8]

Locke: [*dryly*] Most instructive. Master Frege, I see, must have been a Cambridge man – a Platonist.

Socrates: [*roars with laughter*] Plato would have been proud of him! [*He chuckles*] They all suffer from abstract objectivitis.

Alan: What's abstract objectivitis, Socrates?

Socrates: [*chuckles*] It is a disease of the intellect that consists in thinking that if some expression looks like the name of a concrete object only isn't, then it must be the name of an abstract object. Haha!

My friends, you must get it into your heads that thoughts are not *objects* of any kind. They are what is expressed by declarative sentences that *can*, but *need not*, be used for making an assertion. But *what* you assert when you say that things are thus-and-so is not a kind of object. To think a thought is not to grasp an object. Thoughts are not psychological objects, and they aren't abstract objects either, because they aren't objects!

Paul: I agree. But let me just finish filling you in about Frege's ideas. He did hold that thinking, in its higher forms, was made possible for us humans only if it is presented in language. But he claimed that it was perfectly conceivable that there should exist beings that can grasp the same thought as we do without needing to clad it in a form that can be perceived by the senses.[9]

Alan: Let me get a wee bit clearer. Mr Locke and the linguistic idealists hold that thinking antecedes language, since, on the one hand, language *presupposes* the ideas for which words stand, and, on the other hand, thinking is *operating with ideas*, that is, combining, separating and inter-substituting ideas in the mind. Frege, and linguistic Realists, or Platonists, hold that thoughts exist independently of language, and that thinking is a matter of grasping these entities that stand in sempiternal logical relations that are described by the laws of thought. But they also hold that for us humans, thoughts must be presented in the symbolic forms of language and mathematics.

Socrates: That seems a fair summary, my dear boy. You could of course add the Language of Thought hypothesis advanced by young Wittgenstein and later repudiated, and subsequently advanced by others, such as Fodor. Wittgenstein told me that when he was young, he held that thinking just is a kind of language. A thought, he then argued, is no less a logical picture of possible facts than is a proposition of language, so a thought *just is* a kind of proposition.[10] And he held that a thought must have psychological constituents which correspond to the words of language, constituents that stand in the same sort of relation to reality as words.[11] Of course, he rapidly changed his mind about this when he returned to philosophy. But we discussed the Language of Thought hypothesis before dinner, and need not pursue it further.

Locke: Yet it seems evident, does it not, that a Child can conjoin and separate Ideas long before he can speak? And, contrary to the Views of Monsieur Des Cartes, Brutes have Ideas no less than we Humans.

Frank: And cognitive scientists would argue that information-processing presupposes the possession of pre-linguistic concepts.

Socrates: But, my dear friends, we have already been over most of this ground. Combining ideas is not thinking, and it is surely obvious that perceiving is not processing information if 'processing information' is a cognitive or logical operation on propositions with a sense. If you are sensible, you will ask yourselves what a pre-linguistic concept might be, and what would show that a creature had such concepts.

Frank: Well, isn't it enough if the creature can discriminate between things that have a certain property and things that don't?

Alan: Or, more generally, isn't it a sufficient condition for a creature to possess a concept that it can distinguish between things that fall under the concept from things that do not? I mean, doesn't it suffice for a creature to have a given concept that it have an appropriate recognitional and discriminatory ability? And both animals and human babies have a multitude of recognitional and discriminatory abilities.

Paul: No, that's quite wrong. One can teach a bird, or a baby, to differentiate between red buttons and buttons of other colours. But does that mean that they possess the concept of red?

Frank: Why not? If they both possess colour vision, and have been taught, by means of training and rewards, to respond differentially to red things, then why should we not describe them as possessing the concept of redness? Or at any rate, the proto-concept?

Paul: You're both forgetting that the meaning of a word can be described as the place of the word in the grammar of the language. So to know what a given word means is to know its logical powers. To grasp a concept is to know its logical implications, its compatibilities and incompatibilities, and so forth. So if a creature possesses the concept of red, it must know that red is a colour, that anything that is red is coloured. Can a creature have mastered the concept of red without knowing that red is darker than pink?

Alan: I don't follow you, Paul. Surely a child learns how to use the word 'red' before it learns the use of the word 'colour'?

Paul: Perhaps. But in so far as that is true, it has a very poor grasp of the concept. To possess a concept is to possess a linguistic ability, and obviously abilities come in degrees. One may have only partial mastery of a concept, but if one has no more than the ability to distinguish red things from things that are not red, independently of the use of a word that means red, there is no warrant for ascribing possession of the concept of red.

Alan: That sounds reasonable. But what about red being darker than pink?

Paul: Well, if you – a competent, mature language-user – know that the roses in the front rose-bed are all red, and the roses in the rear rose-bed are all pink, then you surely don't need to look to see whether the roses in the front rose-bed are darker than those in the rear rose-bed do you?

Alan: [*hesitantly*] No … No, of course not.

Paul: Right. So you must know that red is darker than pink. And what you know, when you know that, is an inference-rule, namely that from 'A is red' and 'B is pink', one can infer 'A is darker than B'. That inference-rule is partly constitutive of what it is to be red. Similarly, if one rose is red, another is orange, and the third is yellow, then you know that the first is more similar in colour to the second than to the third.

Alan: Ay.

Paul: That is because adequately to have mastered the concept of red is to know that red is more like orange than it is like yellow. And that too is the expression of a rule of inference.

Alan: You mean that if you possess the concept of redness – as a mature speaker of the language does – you must have grasped these inference-rules.

Paul: Of course. These are *constitutive* of being red, not mere *consequences* of being red. And that is part of the reason why mere possession of a recognitional ability is not sufficient for concept possession. Moreover, in many cases, it isn't even necessary. You know what a thousand-sided polygon is, but you certainly can't recognize one on sight. You certainly possess the concept of being old, or of being a fake, but that does not mean that you have the ability to identify an Old Master or a fake on sight.

Alan: I can see that. But it's still open to thinkers like Chomsky to argue tha' the speed and precision of vocabulary acquisition leaves no real alternative to the conclusion that the child has the concepts available before experience with language and is basically learning labels for the concepts that are already part of his or her conceptual apparatus.[12]

Paul: Alan, you've already forgotten what Socrates showed us before dinner. Words aren't names or labels of concepts. Concepts don't have names, and there is no such thing as a label for a concept. *Some* names *express* concepts, but to express a concept is not to label it.

Alan: Ay, I'm sorry, I should have remembered. But what names *don't* express concepts?

Paul: Well, personal names, like 'Tom', 'Dick' and 'Harry', don't, and neither do many other proper names, like 'Paris' and 'London', or 'Monday' and 'Tuesday'.

Alan: All right. But Chomsky might still insist that learning a language is learning to use words to express concepts that one already possesses innately.

Paul: He might. But would he have any reason for doing so?

Frank: Well, I guess he'd say that it was an inference to the best explanation. He'd argue that the speed of language acquisition in the human child is such that there is no better explanation than the assumption that we are born in possession of innate concepts.

Paul: But Frank, old boy, an inference to the best explanation is coherent only to the extent that the 'best *explanans*' has been given a sense. But it is wholly obscure what *counts* as possessing concepts innately. Do we have any criteria for a neonate's being born with innate concepts – *other than* the phenomenon of rapid language-acquisition. But this was what the hypothesis, the inference to the best explanation, was supposed to *explain* – and now it turns out that the phenomenon is the only explanation we have of the meaning of the hypothesis. That human beings have innate tendencies and propensities to respond to linguistic stimuli, to gesture, voice and intonation contour, is indisputable. But the supposition that they have *actually innate* concepts is a very much weightier hypothesis than that, and it is supported by no empirical evidence whatsoever.

Frank: OK ... OK ... Still, let me try one more point to show that thought is prior to language. Y'know William James, the greatest philosopher of psychology we've ever had in my country, observed that there was good empirical evidence to show that one can think perfectly well without knowing a language. A deaf and dumb kid, James wrote, can weave his tactile and visual images into a system of thought just as rational and effective as that of a language user. James relates that there was a guy named Ballard, a deaf-mute, whose reminiscences of childhood prove that one can think even though one cannot speak.[13] Ballard said that when he was a child, two or three years before he learnt to write, he asked himself the question of how the world came into being. When this question occurred to him, he says, he set himself to thinking about it at length. Now, why isn't that definitive? One *can* think without knowing a language.

Locke: That does, methinks, give Quietus to the Language-mongers.

Socrates: Gentlemen, we have got so far in painting our picture, don't spoil it by a few rash and hasty brush strokes. What would you say if one of our friends, say Xenophon, told you that he remembers that before he was born he enjoyed long conversations with Solon?

Alan: You can't remember something that could never have happened. How can anyone remember doing something before he even existed?

Socrates: Quite so, my dear fellow. So what would you say?

Alan: Well, I suppose I'd ask him what on earth he meant. I'd say that I couldna understand his words.

Frank: I don't see the point of this.

Socrates: Well, the sentence 'I remember that before I was born I engaged in conversation with someone' is unintelligible, is it not?

Frank: Yeah, I guess so.

Socrates: Well, is the sentence that Ballard wrote any more intelligible?

Frank: Y'mean when he said that he could remember that before he could use language, he thought about the origins of the world? Why isn't that intelligible?

[A pause]

Paul: Yes, I see. What would it mean for little Ballard to think this?

Frank: Well, exactly the same as what it means for you or me to think about the origins of the universe.

Paul: I think not. What you can be said to think is what you *can* express even if you don't. If I say 'A penny for your thoughts', you can tell me. But what could little Ballard do? He couldn't speak, and he hadn't yet mastered sign-language, let alone learnt to write.

Locke: But if, by fell Fortune, I were bereft of the Power of Movement, I too would not be able to relate my Thoughts on Request.

Alan: But we know what you'd have to do in order to express what you thought. And we know that if not for your paralysis, you'd be able to do it. After all, y' know how to speak, only you can't move your lips and tongue.

Paul: Yes. Saying what you think is within your *behavioural repertoire*. If you were to recover from the stroke that afflicted, you *would* say what you think. But Ballard is not prevented from exercising his linguistic abilities, he does not have any. There is nothing he could do, nothing within his behavioural repertoire, that would *count* as an expression of thoughts about the origins of the world.

Frank: But one *can* think without saying anything either out loud or to oneself. Thought is possible without language.

Alan: Frank, y're confusing two quite different claims. We've seen that it makes sense to say that a person thinks or is thinking something or other, even though he is not using language – not saying anything aloud or to himself. But we haven't established, and I fancy canna establish, that a person can be said to think something that he could not express if unimpeded.

Paul: Good, I think we're getting there. Can we say that the limits of thought are the limits of the possible behavioural expression of thought? Of course, the behaviour may be non-linguistic as well as linguistic, but non-linguistic behaviour is *exceedingly* limited in its thought-expressive possibilities.

Locke: Master Paul, that sounds most ingenious. But I have not fully comprehended your Words. I should be greatly in your debt were you to elaborate somewhat upon that Idea.

Paul: My proposal is that although, as we have agreed, we can think without saying anything, either to ourselves or out loud, what thoughts it *makes sense* to ascribe to us is constrained by what we *could* say out loud, if we so chose and were not impeded either externally or internally. The limits of thought are the limits of the possible expression of thought.

Frank: OK. that's very fancy. But I'm still not convinced by your whole approach. Y' wanna say that thinking isn't the same as speaking, but it's almost like it. I don't understand that. Thinking something isn't at all like talking. For one thing, you can think much faster than you can say anything, aloud or to yourself. Hell, don't all the poets express their amazement at the speed of thought? Y'can think of the solution to a complex problem at a stroke, and then it may take you half an hour to spell it out.

Socrates: No. Now you are blundering around like a mule in a pottery shop, Frank. You need to slow down. We didn't establish that thinking is like talking. What we have now established is that the limits of thought are the limits of its *possible* expression. That correlates what you *can* think with what you *can* express in your behaviour, including your verbal behaviour.

Locke: Nevertheless, the Consideration that Master Frank has mooted is of no small Consequence. Every Man's Experience convinces him that he can think in a Flash. Is the Speed of Thought not a Mystery?

Socrates: Master Locke, you must not confuse what is wonderful with what is mysterious. It is wonderful that nature has endowed us with the power to see the solution to complex problems in a flash. But it only seems mysterious when you misconstrue the expression 'the speed of thought'.

Frank: I don't follow you, Socrates. That's just what *is* mysterious. How on earth can thought be so fast?

Socrates: Frank, we often see, or think we see, the solution to a problem in a flash. But we misconstrue this phenomenon if we think that this is a matter of whizzing in our minds through the manifold steps that lead from the statement of the problem to its solution, and then spelling it out slowly in words. But it's not like emptying an hourglass of water at a stroke as opposed to letting the water drip slowly through in an hour.

Locke: You must say more, Master Socrates, if you wish your Arrows to hit their Targets.

Socrates: Surely it is evident that to see the solution to a complex problem in a flash is the sudden *realization* that one can do something, not the high-speed *execution* of what one realizes one can do. The sudden flash of inspiration, the *Eureka* experience, is a pointer, not a product. Whether one is right to think that one has 'got it', that one really can prove the theorem, resolve the conundrum, solve the problem, remains to be seen. We do, after all, sometimes think that we have seen the solution in a flash, only to discover, when we try to spell it out, that we were mistaken.

Frank: Y'mean that we may think we've twigged it, and then we find that we're wrong.

Paul: Yes. But of course, we don't advertise those cases. We tell tales of Archimedes, but consign to dust the hundreds of village Archimedes who've jumped out of their baths with a 'Eureka', only to find that they did *not* actually have the answer to the problem.

Socrates [*chuckles*] Good.

Alan: So seeing the solution to a problem in a flash is more like the dawning of an ability than like its exercise.

Socrates: Yes, and that is why it is wonderful, but not mysterious. We can, sometimes, realize that we can do something, but realizing that we can solve a problem is not running at high speed through the reasoning that leads to its solution.

Alan: I think I follow you, Socrates, but there is still something that I canna understand. Where does all this leave animal thinking? Surely, animals can think. But *they* can't express their thoughts, can they? So are we going to follow Descartes and *deny* that animals can think?

Locke: Methinks Master Alan has asked the proper Question. They must needs have a penetrating sight, who can see that Dogs or Elephants do not think, when they give all the demonstration of it imaginable, except only telling us, that they do so.[14] Brutes have Senses, no less than Man, and their Senses furnish them with Ideas.

Paul: To be sure, they see and hear, rather better than we do. They suffer pain no less than we do. But does it follow that they can think?

Locke: That is not easy to determine. I imagine that they have not in any great degree the Faculty of comparing Ideas one with another. It seems to me to be the Prerogative of Humane Understanding, when it has sufficiently distinguished any *Ideas* to cast about and to compare them. *Beasts compare* not their *Ideas*, farther than some sensible Circumstances annexed to the Objects themselves. Nor do they, I suppose, compound *Ideas to* make new complex *Ideas*. Moreover, I think, I may be positive in, That the power of *Abstracting* is not at all in them. The having of general *Ideas* is that which puts a perfect distinction betwixt Man and Brutes. For it is an Excellency which the Faculties of Brutes do by no means attain to.

Paul: I see. And if by *lacking general ideas* you mean that they have no concepts, then I agree. For, to be sure, animals do not speak, do not use words and cannot master the use of words. But does that imply, in your view, Mr Locke, that animals can neither think nor reason?

Locke: No, that I cannot concede. If they have any *Ideas* at all, and are not bare Machins – as Monsieur Des Cartes and his followers would have them – we cannot deny them to have some Reason. It seems as evident to me that they do some of them in certain Instances reason, as that they have sence. But it is only particular *Ideas*, just as they receiv'd them from their Senses. They are the best of them tied up within those narrow bounds, and *have not* – as I think – the faculty to enlarge them by any kind of *Abstraction*.[15]

Socrates: Thank you, Master Locke. That's very interesting and thoughtful. Now let's see whether and how we can accommodate these penetrating observations in our reflections.

Paul: Well, for one thing, we surely need to abandon the supposition that thinking is an operation or operations on ideas. On the other hand, Mr Locke is surely right to draw our attention to the generality of concepts and of words that express concepts, and to combination and separation by means of predication and negation, as well as sentential combination by means of operators such as conjunction, disjunction and negation. Not to mention generalization by means of such quantifiers as *All* and *Some*, *Any* and *Many* and so forth. One surely cannot ascribe such essentially intellectual operations to non-language users.

Frank: Well, I'm kinda feeling out of my depth here, but it seems to me that Mr Locke is right to say that animals think and reason in a primitive kind of way. And a great deal of scientific work is being done on animal reasoning, y'know. It's pretty much accepted by respectable animal ethologists that chimpanzees develop a theory of mind that enables them to recognize the beliefs of other chimpanzees and to form intentions and plans to deceive them.

Paul: Well, that's pushing it.

Frank: No, honestly – there's a mass of evidence from chimpanzee behavior that involves hiding food in pretty complicated ways that presuppose that the chimpanzee, who's keen on protecting his hoard of bananas, forms beliefs about the beliefs of the other chimps in his group. I kid you not – they're pretty sophisticated apes, y'know.

Paul: [*rather irritated*] Not that sophisticated, Frank. Ascribing a theory of mind to a chimpanzee that can't even talk is preposterous.

Frank: [*rather annoyed*] I don't see why. And I don't see with what right you can so easily dismiss serious scientific work done by animal behaviorists and ethologists.

Socrates: Now calm down, you two, and try thinking, instead of getting excited and swapping intuitions. The first matter we must settle is a set of agreed cases that exemplify an animal's thinking something to be so. Then we can go to work.

Paul: Well, I am willing to go along with the idea that if I take my dog's leash off the peg and rattle it, the dog will think it is going to be taken for a walk. And if a dog chases a cat that jumps up a tree, and the dog barks excitedly at the bottom of the tree, even though the cat has disappeared, then it thinks that the cat is in the tree.

Alan: Ay, I'll go along with that.

Socrates: But would either of you say that the dog *has the thought* that it is going to be taken for a walk, or that the cat is still in the tree?

Alan: [*hesitantly*] … Well, I'm not sure. It sounds odd!

Frank: That it sounds odd doesn't mean that it isn't true. Maybe it sounds odd because you never have reason to say such things. Perhaps it's too obvious to be worth saying, and that's why it sounds odd to you.

Paul: Hmm. That's the Grice gambit.[16] But I wonder whether it is right. It's a very popular move these days, and is usually wrong.

Socrates: Let us investigate. What is the difference between thinking, or believing, that something is so, and thinking or believing the thought that something is so?

[*A short silence*]

Paul: Yes, that *is* a good question, Socrates. I suppose that *to think the thought that something is so* is to believe the proposition that things are so.

Alan: Ay, and to believe the proposition that things are so is to believe the proposition to be true. But y'canna believe propositions to be true unless you have the concept of a proposition, as well as the concepts of truth and falsehood. And surely no one wants to ascribe such concepts to non-language users!

Frank: [*thoughtfully*] Yeah, that's neat.

Locke: It does, in truth, confirm my Judgement that mere Brutes abstract not, and are wanting in general Ideas.

Frank: OK, Alan. But y'do concede that animals think and believe. So you accept the claim that when I rattle the dog's leash and it rushes to the door wagging its tail, it believes that it's going to be taken for a walk.

Socrates: You must take this slowly, Frank, for the path branches here, and you must take care to follow the correct road.

Frank: I don't follow you. The dog believes it's going to be taken for a walk and that's why it rushes to the door wagging its tail.

Paul: Oh, I see what you mean, Socrates. The dog's belief is not its *reason* for rushing to the door and wagging its tail. Our reluctance to ascribe thoughts and beliefs to animals is not only because that would commit us to ascribing concepts of truth, falsehood and of a proposition to a non-language user; it is also because it makes no sense to ascribe to the animal a reason for wagging its tail and pawing the door, namely: the belief or thought that it is going to be taken for a walk. It is not the belief that explains the behaviour, but rather the behaviour that gives sense to the ascription of believing.

Alan: I don't think I understand that, Paul. Could y'not elaborate a wee bit?

Paul: Well, what I am suggesting is that the significance of denying beliefs, as opposed to believings, to animals is much deeper than at first sight it appears to be. Surely there is no such thing as an animal believing something to be so in advance of acting, and reflecting on whether things' being so is a good enough reason for acting or whether it is outweighed by other reasons. I know what it is for us to weigh reasons and come to conclusions, but what on earth would it be for a cat or a dog to do so? Just think how many things they would have to be able to do before it could be said of them that they weighed reasons pro and con, and on reflection came to such-and-such a conclusion.

Locke: I warrant that you are in the right, Master Paul. For only if one can *reason*, can one act *for Reasons*.

Frank: Well, I don't see that. Look, a cuckoo lays an egg in another bird's nest. The bird doesn't notice, and when the fledgling hatches, the bird believes it to be her own. And that's why she doesn't throw it out of the nest. So she feeds it. Isn't that acting for a reason?

Paul: No, Frank. The female bird *treats* the fledgling as her own. It's not the belief that is her reason – we've agreed that it is misguided to ascribe a belief to her. Nor is it her believing that the fledgling is her own that *explains* her behaviour. On the contrary, it is the ascription of believing that redescribes the behaviour.

Frank: That's just not clear to me. What about this case. Y'know that a vervet monkey when it's attacked by another troop emits a leopard-warning cry in order to get the attacking group of vervet monkeys to scatter. Now that seems to me to suggest very clearly a piece of reasoning. The monkey possesses the information that a leopard-warning cry will frighten the attacking monkeys and get them to flee, and it wants to get them to stop attacking it, so it reasons that giving the leopard-warning cry will get them to flee, and that's why it acts.

Alan: But surely, Frank, we don't speak of animals *possessing information*.

Frank: Well, if we don't, I'd like to know why not. Dammit, they know things, don't they?

Alan: Ay, the dog knows the way home from the park, and it knows its master. And your monkey knows leopards are dangerous – it responds to the sight of a leopard with fear and flight, and it emits leopard-warning cries. But it doesn't follow that it possesses *information* about leopards. It hasn't learnt the truth of various propositions about leopards, it has learnt how to respond to leopards.

Paul: That's right. Look, Frank, what substance can we give to your cognitive and cogitative suppositions? In the absence of a language, what would it be for the monkey to reason thus?

Frank: Well, what substance is there in our case?

Alan: Oh, Frank. We've already been over that. We can give justifications for what we are about to do, and give justifications for what we have done. We can reflect on our various reasons, and weigh them before deciding what to do.

Frank: So how would you describe what the vervet monkey is up to?

Paul: I don't think there is any difficulty. Its behaviour is purposive and is also a consequence of past experience. But that does not mean that it involves possession of information, reasoning and foresight. A vervet's cries are innate – innate responses to something moving on the ground below or flying in the sky above. With time its cries become more discriminatory: it responds to eagles, but not to crows, and to leopards and not to other animals on the ground. Obviously it may have learnt, presumably by chance, that if it gives a leopard-warning call, all the monkeys scatter. So it acts in order to get the attacking monkeys to scatter. Its behaviour is purposive all right, but not reasoned. Its cry is like an instrument that causes the other monkeys to scatter, but it means nothing and the monkey does not mean anything by it.

Frank: OK. I'm not sure now … Let me try one more time. Y'know that a lot of research has been done on chimps deceiving each other. I mean, if a chimp gets a banana, it'll try to hide it from the others who might take it away. If it knows that the alpha male is watching it when it is hiding the banana, it'll wait until the attention of the alpha male is distracted, and then move the banana to another hiding place. Now many animal behaviorists hold that this shows that chimpanzees have a developed theory of mind that enables them to predict the beliefs of other chimps, and to adjust their own behaviour to the beliefs of other chimps.[17]

Socrates: [*laughs long and heartily*] That is wonderful! [*He wipes his eyes*] I don't have a theory of mind myself [*he breaks out laughing again*], and your friends think that apes do!

Frank: I don't see what's so funny. And it seems to me that you do have a theory of mind, as we all do. Hell, we know that when someone hurts himself and cries out, we can infer that he is in pain, and that when someone turns white and trembles in circumstances of danger, we can infer that he believes that he is in danger and that he is frightened. We know that if y'say something offensive to someone, and he turns red and shouts at you, then he's angry. Why isn't that a theory of mind? It's a theory that enables you to predict what others are feeling and what they believe on the basis of their behavior. We can't *see* what's going on in their mind, but we can observe their behavior. So we can infer from their behavior and from our theory of mind what they're thinking and feeling. That's what you guys call 'an inference to the best explanation' isn't it?

Locke: Master Frank, I must venture to say that it is incumbent upon you to explain how Brutes that cannot speak and can form no general Ideas can nevertheless construct an abstract Theory. For Beasts make no Propositions and reason not at all.

Socrates: My friend, your examples show that you understand the words 'pain', 'fear' and 'anger'. For that you need no theory. You need only to have learnt how to use these words. To judge a man to be in pain when he has injured himself and cries out is not to make an inference to the best explanation, as you call it. Indeed, it is not even to make an inference. You can *see* that the poor devil is in pain. And the pain-behaviour warrants your judgement.

Frank: Y'mean that you're a behaviorist? That pain is just pain behavior?

Paul: No, no, Frank. You're rushing the fences. We mean no such thing. Pain-behaviour is constitutive evidence for pain. It is not inductive evidence, nor is it the grounds for an hypothesis. To learn how to use the word 'pain' is to learn that pain behaviour in appropriate circumstances warrants the application of the word, even though you do not *infer* that the man is in pain from his pain-behaviour. You may infer that your wife is in pain if you see that she has taken some analgesics, but you don't *infer* that she has a headache if she holds her head with a grimace of pain and says 'Oh, I've an awful headache'. Moreover, to learn the use of the word 'pain' is also to learn to replace one's natural pain behaviour with pain utterances. Instead of crying out or moaning, you learn to say 'It hurts' or 'I have a pain'. There is no *theory* here, only a linguistic practice.

Frank: But then how d'you explain the behavior of chimps concealing their food from the alpha male? Hell, the animals are highly sensitive to the beliefs and desires of the other chimps. That surely requires that they can read their minds, and to read their minds they need a theory of mind.

Paul: No, no, Frank. You've got hold of the wrong end of the stick. The chimps are sensitive to the behaviour of other chimps. They learn to read their behaviour, not to read their minds. They learn to respond to the wants of other chimps only in so far as they learn to respond to their desire-manifesting behaviour. Similarly, they learn to respond to the cognitive behaviour of other chimps, not to impute beliefs to them.

Frank: Well, I'm not sure about that. I mean …

Locke: Gentlemen, it is with Reluctance that I must leave you. The play at the New Globe will shortly commence, and I do not want to be late.

Socrates: What, has Will Shakespeare put on a new play?

Locke: That is so. It is a subject that will gain your interest. It is entitled 'Alcibiades – a Tragedy'.

Socrates: That is indeed a subject for a great tragedy. A much better subject than his 'Coriolanus'. We must see it. Gentlemen, enough philosophy for this evening. If we don't want to miss the play, we must stop.

[*They all rise from their chairs and walk off together through the trees*]

Notes

1   J. Hadamard, *The Mathematician's Mind: The Psychology of Invention in the Mathematical Field* [1945] (Princeton University Press, Princeton, NJ, 1996), p. 142.

2   R. Penrose, *The Emperor's New Mind*, rev. ed. (Oxford University Press, Oxford, 1999), pp. 548–49.

3   F. Galton, quoted by Penrose, *The Emperor's New Mind*, p. 548.

4   J. Locke, *An Essay Concerning Human Understanding* [1690], III-ii-1.

5   A. Damasio, *The Feeling of What Happens* (Heineman, London, 1999), p. 107.

6   G. M. Edelman and G. Tononi, *Consciousness: How Matter Becomes Imagination* (Allen Lane, Penguin Press, London, 2000), pp. 215–16.

7   G. Frege, *Posthumous Writings* (Blackwell, Oxford, 1979), p. 270

8   Ibid., p. 145.

9   Ibid., pp. 142, 269.

10   Wittgenstein, *Notebooks 1914–16* (Blackwell, Oxford, 1961), under 12. 9. 16.

11   Wittgenstein, letter to Russell, 19. 8. 1919, repr. Wittgenstein, *Notebooks*, p. 129.

12   N. Chomsky, *Language and the Problems of Knowledge* (MIT Press, Cambridge, MA, 1988), pp. 27–28.

13   W. James, *The Principles of Psychology*, vol. I (Henry Holt, New York, 1890), pp. 266–67.

14   Locke, *An Essay Concerning Human Understanding*, II-i-19.

15   Ibid., II-xi-11.

16   Paul Grice introduced a variety of pragmatic conversational principles by means of which he defended the view that, for example, whenever we see, we seem to see, but it is not worthwhile saying so, since it is too obviously true. Similarly, he suggested, whenever we do something, we try to do it and succeed, but we don't say that we tried because it is too obvious. This analysis has not gone unchallenged.

17   See, e.g., Frans de Waal, *Primates and Philosophers* (Princeton University Press, Princeton, NJ, 2006).

SUPPLEMENTARY READING

P. M. S. Hacker, *The Intellectual Powers: A Study of Human Nature* (Wiley-Blackwell, Oxford, 2013), chap. 10.

Bede Rundle, *Mind in Action* (Clarendon Press, Oxford, 1997), chaps 3–4.

Gilbert Ryle, *Collected Essays*, vol. II (Hutchinson, London, 1971), essays 19, 22, 30, 34, 36.

Gilbert Ryle, *On Thinking* (Blackwell, Oxford, 1979), essays 1–5.

SECTION 5

A DIALOGUE ON
OWNERSHIP OF PAIN

INTRODUCTION

The thought that another person might have my pain seems absurd. You can, it seems, have an exactly similar pain to my pain, but you cannot have the very same pain. As the great nineteenth-century logician Gottlob Frege wittily put it, 'You can't have my pain, and I can't have your sympathy.' In short, yours is yours and mine is mine – another man's pain is another pain. Peter Strawson, in chapter 3 of *Individuals* (1959), held that experiences in general (and pains in particular) are privately owned and inalienable. Indeed, they are not merely inalienable, but *logically* inalienable. I can't give you my pains, and you can't have them, even if you wanted to. Each human being has his own 'world of consciousness', as John Stuart Mill said, and only the owner of the world of consciousness can enter it and observe what is there to be found. What is *in* my world of consciousness *cannot* be in yours. Nor can it emigrate from my world of consciousness with an entrance visa into yours. Each subject has his own 'world of consciousness' to which he has privileged access. For no one can enter the world of another's consciousness. Conscious states, as John Searle has put it, are *ontologically subjective*. They have a subjective form of existence, for they exist only when they are had by a conscious subject, animal or human. Obviously, there are no pains without sufferers – one cannot find pains floating around the room unowned. Experiences have to have an owner. Every experience is someone's experience. That, it seems, is a profound metaphysical truth.

This, roughly speaking, is what virtually all philosophers have thought. It is, indeed, the natural way to think. But as should be evident from previous dialogues, in philosophy the natural way to think is all too commonly the wrong way to think. The first philosopher, as far as I know, to challenge received wisdom, and these received distinctions, was Wittgenstein. He did so in his great work the *Philosophical Investigations*, §§253–54, but with a brevity that makes excessive demands upon his readers. 'What pains are "your pains"?', he queries:

> What counts as a criterion of identity here? Consider what makes it possible in the case of physical objects to speak of 'two exactly the same':

for example, to say, 'This chair is not the same as the one you saw here yesterday, but it is exactly the same as it.'

Then he tantalizingly adds:

In so far as it makes *sense* to say that my pain is the same as his, it is also possible for us both to have the same pain.

And so he leaves us. When he wrote in the preface to his book that he would not like his writing to spare other people the trouble of thinking, he really meant it.

I first wrote about this passage and the view expressed by it in *Insight and Illusion* (1972). What I wrote conformed to the conventional view of the matter that different people may have qualitatively identical pains, but not the numerically identical pain. Indeed, I interpreted Wittgenstein as expressing that view too. It took me more than a decade, and the assistance of my friend Norman Malcolm, to realize that this was the exact opposite of the truth. When I came to write a second edition of *Insight and Illusion* (1986), I rectified my error. In the book of commentary that I wrote on these passages, *Wittgenstein: Meaning and Mind* (1990, revised edition 2019), I offered detailed exegesis of his remarks, as well as detailed exposition of the issue in an essay entitled 'Privacy' in the first edition, which I split into two new essays in the revised edition 'Only I can have' and 'Only I can know'. Although I may have persuaded some readers that this was what Wittgenstein meant, much to my surprise, I have persuaded hardly anyone that what he meant was also correct. I remained, and still remain, convinced that he was right. But it is true that this tiny little problem is one of the smallest and most tightly drawn knots in the whole of philosophy. And, of course, the more tightly drawn the knot, the harder it is to untie it. Nevertheless, the problem is by no means a trivial curiosity, for it ramifies into far-flung corners of philosophy of psychology. If one does not get this right, one will make substantial and deep mistakes in domains far removed from worries about whether two people can have the same pain – in psychology, cognitive science and cognitive neuroscience.

While it is indeed a tiny problem, the difficulty of resolving it is formidable. Although the road to the solution is straightforward, there are mountains of misconceptions that stand in the way of finding the right road. Or, at any rate, so it seems to me – and this is what I have tried to show in the following dialogue.

The dialogue includes three great dead philosophers, Frege, Strawson and Wittgenstein, the former two adopting the received view that experience is privately owned, and the latter combating it. In addition, there are two distinguished living philosophers, John Searle and Wolfgang Künne. They too defend the received view. Searle has written extensively, with his customary clarity and panache, about the essential subjectivity of experience. His views are always worth careful consideration. My friend Wolfgang Künne generously contributed an essay to a *Festschrift* dedicated to me, *Wittgenstein and Analytic Philosophy* (2009), in which he gently castigated me for my heresies. He was supposed to have come to the symposium for which I originally wrote this dialogue. So I transported his shade into Elysium to participate in the dialogue, hoping that we should be able to continue to discuss the matter in our sub-lunary world at the symposium. Alas, he was unable to come – but I hope he finds it an intellectual entertainment.

Eighth Dialogue

CAN YOU HAVE MY PAIN? CAN DIFFERENT PEOPLE HAVE THE SAME PAIN?

Protagonists:

Ludwig Wittgenstein: the greatest philosopher of the twentieth century; a shock of greying hair swept to his left temple, piercing blue eyes; an English accent with occasional Germanisms; dressed in grey slacks with a sports jacket and open-necked shirt.

Gottlob Frege: the greatest logician of the late nineteenth and early twentieth century; wears a beard, is dressed in a dark formal suit and wears bow tie; a thick German accent.

John Searle: a distinguished Californian philosopher from the University of Berkeley; small of stature, with dark eyes and black hair; casually dressed in blue jeans, with an open-necked shirt and jacket; a broad Californian accent and loud voice.

Wolfgang Künne: a distinguished German philosopher from the University of Hamburg; balding, wears glasses; dressed in slacks and sports jacket with a tie; speaks almost perfect English with a slight German accent.

Peter Strawson: sometime Waynflete Professor of Metaphysics at the University of Oxford; balding; dressed in sports jacket and wearing a tie; a deep voice and slightly clipped manner of speech; speaks slowly and meticulously.

The setting is an Oxbridge Common Room in Elysium, where the shades of the dead and of the living are conversing after dinner. The French window is open to a great lawn, and the stars are just beginning to show in the twilight. All except Wittgenstein are seated in comfortable, dark brown leather armchairs, around a low table on which their post-prandial drinks are placed. Wittgenstein is seated on a wooden chair, drinking a glass of water. Frege is smoking a pipe. There are some books on the table, some of them open.

Wittgenstein: You know, philosophical problems really are like knots we tie in our understanding. The smaller the knot, the harder it is to unravel. If you tug on the thread, the knot gets smaller and smaller – and it becomes the very devil to untie.[1]

Frege: *Ja*, Wittgenstein; we all know that you spin lovely metaphors and similes. What exactly do you have in mind?

Wittgenstein: Well, actually, I was thinking of something you wrote?

Frege: So! And what displeased you?

Wittgenstein: Oh no – what you wrote did not *displease* me. In fact, it intrigued me. Language lays the same traps for us all, but we do not all fall into them. If you, of all people, fell into this one, then it must be very hard to avoid. In your paper, 'Der Gedanke' ('Thoughts'), which, you may remember, I didn't like at the time, you wrote:

> Even an unphilosophical man soon finds it necessary to recognize an inner world distinct from the outer world, a world of sense impressions, of creations of his imagination, of sensations, of feelings and moods, a world of inclination, wishes and decisions.[2]

and you went on to call the inhabitants of this 'world' of yours 'ideas'.

Frege: Ach, Wittgenstein – it is not just *my* world. Everyone has such a world. It is the world of consciousness. And what this world contains are *objects* that are of a mental nature. In this I am in agreement with your John Mill.

[*He leans forward and picks up an open copy of Mill's* An Examination of Sir William Hamilton's Philosophy *from the table*]

Here is what he wrote:

> I observe that there is a great multitude of other bodies, closely resembling in their sensible properties ... this particular one, but whose modifications do not call up, as those of my body do, *a world of sensations in my consciousness.* Since they do not do so in my consciousness, I infer they do it out of my consciousness, and that to each of them belongs *a world of consciousness* of its own, to which it stands in the same relation in which what I call my body stands to mine, ... Each of these bodies exhibits to my sense a set of phenomena (composed of acts and other manifestations) such as I know in my own case, to be effects of consciousness, and such as might be looked for if each of the bodies has really in connection with it a world of consciousness.[3]

So Mill. Now surely, what he says is undeniable! I disagree with John Mill about the nature of logic and of number, but I am in complete agreement with what he says here.

Searle: Yeah! I think that John Stuart Mill and Frege here are right. But I think that we can now put it more precisely. I don't much like this talk about *worlds* – I think that there is only *one* world. As Virginia Woolf said, 'One of the damned things is quite enough'. But there are things in the world belonging to different ontological categories. There are objective things, and there are subjective things. And by this I don't mean epistemic modes – as when we distinguish between subjective opinion and objective fact. What I mean are *ontological categories*. Now what Frege calls 'ideas' are in fact *conscious experiences* or *conscious states*. What I am talking about, and what you [*he turns to Frege*] and Mill were talking about, is *ontological subjectivity*. What characterizes all conscious phenomena is subjectivity. Consider, for example, the statement, 'I now have a pain in my lower back'. That statement is completely objective in the sense that it is made true by the existence of an actual fact and is not dependent on any stance, attitudes or opinions of observers. However, the phenomenon itself, the actual pain itself, has a subjective mode of existence.[4]

Künne: John, I really don't understand what you mean by 'subjective mode of existence'.

Searle: OK, let me put it differently. We're talking about conscious states. All conscious states have a qualitative character that is experienced by the person that has them – by the subject. Conscious states are ontologically subjective. Consciousness has a first-person ontology. It exists only as experienced by a human or an animal subject and in that sense exists only from a first-person point of view.

Wittgenstein (*indignantly*) **:** A first-person point of view? What is a first-person point of view? When you have a pain, do you have it from a point of view? I know what it is to express a judgement from a moral, or a political, or an economic point of view – or from my own point of view, when I give my opinion. But what it is to have a pain from a point of view!!

Strawson: [*trying to calm him down*] Wittgenstein, I sympathize with your objection to the phrase – but it really doesn't matter, and it will divert us from our theme – which is what Searle is calling 'ontological subjectivity'. What is your reaction to what Searle said, Frege?

Frege: I think that what Searle is saying is very similar to what I wrote in my old paper that Wittgenstein didn't like. Let me put it *so*: physical objects are actual and objective. Abstract objects are no less objective – they are real, but non-actual. But objects in the inner world are quite different. They are subjective – as Searle says. I wrote, 'It seems absurd to us that a pain, a mood, a wish, should go around the world without an owner, independently. A sensation is not possible without a sentient being. The inner world presupposes somebody whose inner world it is.' Maybe Searle is right that talk of worlds here is misleading. What I meant was that, for example, one cannot find pains in the room – only people in pain. One cannot find hopes or fears in the cupboard – only cups; or perceptual experiences in the street – one can only *have them* in the street. What I meant by 'inner objects' are conscious experiences that essentially *belong* to someone. Ideas or conscious experiences are *had*. So, for example, one does not perceive one's perceptual experiences – one *has* them. *What* one perceives are what one's experiences are experiences *of*, but one *has* the experiences. Ideas or experiences are *essentially owned*. Things of the outer, physical world, are essentially independent.

Searle: Yeah, that's what I meant when I said that subjectivity is an *ontological* category. Experiences have a subjective mode of existence. They exist only when they are experiences had by a human or animal subject. In this respect they differ from nearly all of the rest of the universe, such as mountains, molecules and tectonic plates, which have an objective mode of existence.

Frege: *So.* Now, such inner objects are not only owned, they have only *one* owner and they cannot change ownership. I cannot transfer my ideas to you. One cannot bring together in one consciousness a sense-impression belonging to one consciousness and another sense-impression belonging to another consciousness. An object within one inner world – such as my pain – can no more enter another inner world than it can escape to the outer world.

Now, just as mental objects need an owner, so too, they cannot have more than one owner and cannot change owners. That is why I wrote 'Nobody else has my pain. Someone may have sympathy with me, [*he chuckles*] but still my pain belongs to me and his sympathy to him. He has not got my pain, and I have not got his feeling of sympathy.' [*He laughs*] Not only can't you have my pain when I have it, you can't have my pain even when I cease to have it! If I cease to have a pain, then it ceases to exist – the pain has not only passed, it has passed away. [*He chuckles again; the others laugh. Wittgenstein isn't amused*]

Wittgenstein: Frege, do you really not see any difficulty here? Why *can't* I give you my pain? Is it because it is so difficult? Could I *try* to give you my pain?

Frege: *Ach*, Wittgenstein – can't you see that it is not possible. You can kick yourself in the shins and then kick me in the shins – but your pain will be your pain and my pain will be mine.

Wittgenstein: You mean that it is 'metaphysically impossible' to give you my pain?

Searle: Well, I don't know about metaphysically impossible – but it is certainly ontologically impossible. It is just a plain fact that pain has a subjective mode of existence. Now, what does this mean? It means that because of its subjective mode of existence, its existence is a first-person existence. It must be *somebody's* pain; and this in a much stronger sense than the sense in which a leg must be somebody's leg, for example. Leg transplants are possible; in that sense pain transplants aren't. This is a general truth about conscious states. I have a special relation to my conscious states, which is not like my relation to other people's conscious states. And they in turn have a special relation to their conscious states, which is not like my relation to their conscious states.

Frege: *Ja, Ja.* That's right.

Wittgenstein: No, no. It's not right at all! It is not as if we could *conceive* of unowned pains, only we don't come across any. Nor is it as if we can't come across unowned pains because of some limitation in our constitution. We don't even know what it would be like to come across a pain that wasn't someone's pain. If someone were to say 'I found a pain under my bed yesterday', we should have no idea what he meant. What Searle calls 'ontological impossibility' is *not a possibility that is impossible*. So both of you have to explain what this impossibility is. After all, it is not a brute fact that could be otherwise – it seems to be a super-fact, a *meta-physical* fact. And neither of you have explained what that means.

Strawson: Well, I agree that invoking metaphysical or ontological possibilities and impossibilities is unhelpfully mystifying. It is not something I go in for at all. I think that Frege and Searle have not really got to the bottom of the matter. For the issue is not an ontological or metaphysical one. It is, in a generous sense of the term, a *logical* one. It is a matter of the identification and re-identification of particulars.

It is not helpful to say that it is *impossible* for there to be an unowned pain, or that it is *impossible* for me to have Frege's headache – for all that does is to leave us bewildered by the nature of these putative impossibilities. The truth of the matter is that it *makes no sense* to speak of pains that are no one's pains, or to suggest that the identical pain which was in fact one's own might have been another's. We are not up against adamantine impossibilities. Rather, we have arrived at the bounds of sense. I hope you will indulge me a little and allow me read to you what I wrote on the matter years ago in my book *Individuals*, for I don't think I could put it better today.

[*He leans forward and picks up a copy of* Individuals *from the table and pages through it until he finds the passage he is looking for*]

Yes, here it is on page 97. This is what I wrote:

> We do not in fact have to seek far in order to understand the place of this logically non-transferable ownership in our general scheme of thought. For if we think ... of the requirements of identifying reference in speech to *particular* states of consciousness, or private experiences, we see that such particulars cannot be identifyingly referred to except as the states or experiences *of* some identified *person*. States, or experiences, one might say, *owe* their identity as particulars to the identity of the person whose states or experiences they are. From this it follows immediately that if they can be identified as particular states or experiences at all, they must be possessed or ascribable ... in such a way that it is logically impossible that a particular state or experience in fact possessed by someone should have been possessed by someone else. The requirements of identity rule out the logical transferability of ownership.[5]

It seemed to me then, and it seems to me now, that this *logical* account renders it perfectly intelligible that we should say that there cannot be unowned pains, and that another person cannot have my pain, while avoiding the dubious intelligibility of appealing to metaphysical or ontological impossibilities. The apparent impossibility is actually a limitation of *sense* which is not an arbitrary convention, but has its rationale in the requirements of reference and identification. Pains, and experiences in general, are identifiability-dependent upon the persons whose pains and experiences they are.

[*He lights a cigarette, inhales slightly nervously and looks at Wittgenstein*]

Wittgenstein: I see. [*Pause. He puts his face in his hands*]

We must think this through. [*Another pause*]

Now, Strawson has nicely transformed what appeared to be a meta-physical mystery into a logical, or as I would prefer to put it, a grammatical, claim.

Strawson: [*sotto voce*] I wouldn't. I wouldn't.

Wittgenstein: [*continues, without paying attention to Strawson's muttering*] ... a grammatical claim of identity-dependence of pains on owners. He thinks that the identity of pain turns on *whose* pain it is, and therefore another person *can't* have my pain. I may have a headache. If you also have a headache, it seems, we *call* what you have 'another headache', and we do not say 'You have my headache'. This is because the identity of the particular that is your pain depends upon your identity. So I can't have your pain. Now, trying to getting rid of meta-physical necessities is no doubt admirable. But have we put the question marks deep enough down?

Künne: Can't Strawson's argument be strengthened by reflecting on pain-location? It is clearly constitutive of physical pain to have a bodily location. It always makes sense to ask 'Where is the pain?' or 'Where does it hurt?' – and many pains are differentiated by their bodily location. One can't have a headache in one's back or a backache in one's head. The location of pain is obviously *one* criterion of identity of pain. So, if Jack has a toothache and Jill has a backache, then they certainly have different pains, since a backache is a different pain from a toothache. Now, if you grant this much, then you must surely grant that two people can't have the same pain. For Jack's pains are the pains in *his* body and Jill's pains are the pains in *her* body. And Jack's body is different from Jill's body. If there is a pain in Jack's knee, it is Jack's pain. Jill may have a similar pain in her knee. But since her knee is not Jack's knee, the pains are in different places. So it is obvious that two people can't have the identical pain. Difference of location surely implies numerical difference of pains. Their pains may be qualitatively identical, but they can't be numerically identical. They are, as Strawson says, different particulars. They are owned by different people and they have different locations.

Wittgenstein: Hmm. Philosophers should greet each other with the words 'Take it slowly'. We must slow down. [*He takes a sip of water and pauses for thought*] We distinguish, with respect to things like chairs, between the same and being exactly alike. This chair [*he thumps the chair he is sitting on*] is the very same chair I sat on last night when we were talking about logical connectives. It hasn't even been moved from its place. But if it needs repair, it can be replaced with another chair from the same set that is exactly alike. Now, what makes it possible to draw this distinction in this kind of case?

Strawson: Well, it is surely this: material objects consist of matter. The chair you are sitting on consists, let us suppose, of fifteen pounds of mahogany. The other chairs of this set likewise consist of fifteen pounds of mahogany. But of course, the wood from which your chair is made is a distinct *specific quantity* from the specific quantities of wood from which the other chairs are made. Two chairs are material space-occupants of the same general kind, but they consist of different specific quantities of matter that cannot occupy the same place at the same time. But, of course, the chairs may, to all intents and purposes, share all their other properties. So they are numerically distinct, but qualitatively identical.

Wittgenstein: Hmm. That is on the right lines, although the jargon may be misleading. We do distinguish, when we speak of things like chairs, between being the same and being exactly alike. If you wish, you may dress that up in the jargon of numerical identity as opposed to qualitative identity. But that can be misleading.

Strawson: Why so? I don't see anything misleading here.

Wittgenstein: Look! Replacing the expression 'the same' by 'identical' or 'numerically identical' is a typical expedient in philosophy. It makes it look as if we were talking about fine shades of meaning and are just looking for the right words to hit on the correct nuance.[6] But that is not what is going on here at all. What is going on here is the transposition of an element of one language game into another, where it has no place – like introducing a knight piece into draughts.

It is much easier to tempt ourselves that the headache that two different people have is qualitatively the same but numerically distinct than it is to persuade ourselves that they are exactly similar but not the same. We are perfectly familiar with the distinction between being the same and being exactly similar in the domain of things like chairs. We are not familiar with its application to experiences in general or pains in particular. We know what it is for two people to have the same headache, but if we are told that A's headache is *exactly similar* to B's, but nevertheless different – we would not understand what we were being told. If your headache and my headache really *are* exactly similar, we should naturally respond, then there *is* no difference between them.

Künne: Of course, pains are not material things. It is people's bodies that are material things. And two people cannot occupy the same space at the same time, just as two chairs cannot occupy the same place at the same time. Now, isn't there at any rate an analogy with pains? You are sitting over there, and I am sitting here. Suppose you have a pain in your knee, and I have a similar pain in my knee. Then you have a pain in a different place from me! The pains may be exactly alike, that is, to use the jargon you dislike, they may be qualitatively identical, but they are obviously not the same – they are numerically distinct, since your pain is in your knee and mine is in my knee.

Wittgenstein: All right. Let's not quarrel over the jargon. First of all, reflect on how we determine the location of another person's pain? It is clear that the criterion for where a person has a pain is where he *says* it hurts, where he *points* when he's asked where it hurts, and what part of his body he *assuages*. If two people say that their knee hurts, if they both point at their respective knees when asked where it hurts, or if they both rub their respective knees to assuage their pain, then each has a pain in the knee. And now, isn't this what it is for different people to have a pain in the same place? Isn't this exactly what we *call* 'having a pain in the same place'? So isn't it wrong to say that different people cannot have a pain in the same place because their bodies are in different places?

Frege: Now, Wittgenstein, it is you who is going too fast. [*He relights his pipe and puffs on it*] All your argument shows is that in *ordinary language* the phrase 'pain in the *same* location' means the same as 'pain in the *corresponding* location'. But as I wrote many years ago, 'Someone who wants to learn logic from language is like an adult who wants to learn how to think from a child. When men created language, they were at a stage of childish pictorial thinking. Languages are not made so as to match logic's ruler.'[7] A *corresponding* location is NOT the *same* location!

Wittgenstein: Well, I disagree with your attitude towards ordinary language *and* towards logic. But we don't want to go down that road now. Let me show you that the business of location is not the real issue here. There is an easy way to sidestep the matter of location. Suppose two Siamese twins are joined at the knee. If both have a pain at the point of juncture, then they both have a pain in the knee. But *now* it is *the very same knee* to which they point when asked where it hurts. Here one cannot say that twin A can't have the same pain as twin B, because A's pain is in his knee and B's pain is in his knee, for they have the very same knee. So if they describe the pain in exactly the same terms, must you not admit that they have the same pain?

Frege: No! – I don't see that at all. A cannot have B's pain, and B cannot have A's pain. That's the end of the matter. *Another person's pain is another pain!*[8]

Wittgenstein: Good. Very good. [*He pauses for thought, hand on brow*] What your response shows is that the question of pain-location is really a red herring. The decisive point that moves one to deny that different people can have exactly the same pain has nothing to do with the location of pain. It is because, as one insists, *Mine is mine, and yours is yours.* What determines the identity of a pain is above all the *person.* Having a pain, Strawson suggested, is a matter of *logically non-transferable ownership.* So two people cannot have the identical pain, only a similar one. Is that not what you want to argue, Strawson?

Strawson: Yes. That is indeed what I thought, and I am still inclined to think it now. You can't have the same pain as I – the numerically identical pain – because you can't own *my* pain. But, of course, you can have the qualitatively identical pain.

Wittgenstein: Yes. [*He gets up to pour himself another glass of water from the jug on the sideboard, takes a sip or two and stands still, thinking for a few moments*] You all seem to agree on three points. First, that to have a pain is to stand in a relation to what Strawson calls a 'particular'. I won't say anything about the idea of a particular, Strawson, although I have grave qualms about it. Secondly, that this relation is a relation of ownership. And that thirdly, this kind of ownership is logically non-transferable. Now, … where should we begin? … [*He pauses, and then sits down again*]

Let's first enquire into the notion of 'logical ownership'. Ordinary ownership is a *legal relation* between a person and the thing owned. The car I have belongs to me – it's mine. But I might sell it to another. Then it will belong to him – it will be his. So ownership is transferable. I might share my car with another. Then we would both own – both *have* – the very same car. So ownership is shareable.

Now, how do we come by the idea of *logical* ownership of pain, or, more generally, of experience? We think that *having* a pain signifies a logical relation of ownership between the sufferer and his pain, just as *having* a car signifies a legal relation of ownership between an owner and a chattel. Similarly, we think that the car's *being mine* signifies a *legal* relation of possession, and so too, the pain's *being mine* signifies a *logical* relation of possession.

[*He pauses, holding his head in his hands*]

Look! We might say that possession is the *representational form* of experience (and hence too of pain). This is how we *present* pain to ourselves in the grammar of our language, just as we present *meaning something by what we say* in the form of an act, and we present *thinking* in the form of an activity. But surface grammar here is deeply misleading. It is remarkable how the auxiliary verb 'to have' can lead us astray. Someone's having chattels is indeed a relation of ownership between a person and his chattels. But we also speak of *having* a father and siblings, and we speak of 'my father' and of 'my sisters'. This does not signify any relation of ownership, but filial and fraternal relationships. We speak of *having* a promissory obligation. That too is a relation, but not a relation of ownership. It is not a relation between us and what we have, namely an obligation. It is rather a normative relation between us and the person to whom we made the promise. We speak of *having* in hundreds of other different kinds of cases in which *having* does not signify ownership of *any* kind. And in many cases, such as having a sharp tongue, having a sense of humour, having a good mind, having a good time, having a train to catch, not only is no *ownership* is in question, no *relation* of any kind is in question either.

Searle: But Professor Wittgenstein, can't you see that each one of us, each subject of experience, stands in a special relationship to *his* or *her* experience. OK, you convince me that talk of ownership here may be misleading. But still, the relation of every person to their own experience is unique. Now I don't mean that each person has what used to be called *privileged access* to their own experience and only to their own experience. I agree that this is just another can of worms. And I don't mean that each person can *introspect* their own and only their own experiences. That's just another confused metaphor. What I mean is that experiences, and consciousness in general, are essentially subjective. They have a first-person ontology. They exist only as experienced by a human or animal subject.[9]

Wittgenstein: Well, if you mean that sentient beings are sometimes in pain, that when they cut themselves in the hand, their hand hurts – and that trees and stones are not sentient beings, and so there is no such thing as a tree's hurting itself or a stone's being in pain, then of course that is correct. If you mean that when I injure myself, *I* am in pain and you are not, then that too is normally correct. But that does not imply that having a pain is a *relation* between a person and a pain.

To have a headache is for your head to ache – but is that a relation? Between your head and its aching? If you cut your hand, your hand bleeds. Is that a relation between your hand and bleeding?

Searle: But what about conscious experience in general. Isn't it obvious that each person stands in a special relation to his own conscious experiences? The person who has an experience feels the qualitative character of that very experience – and no one else does. That is the subjective character of experience.

Wittgenstein: We shall have to talk more about the qualitative character of experience. But, honestly, not now. Look, to have what you are calling a 'conscious experience' is not to stand in a relation to something called an experience. To experience fear, or joy, or anger is to be frightened, joyous or angry. To have a visual experience is to see – but it is not to *stand in a relation* to seeing. Of course, it may also be pleasant or unpleasant – but to enjoy looking at something is not to stand in any relation to looking. But let's leave all these side-roads for some other evening, and get back to our high-road.

To have a pain is not to stand in a relation to a pain. Pain is not a relatum – it is not a kind of *object* at all. That is why I wrote in my *Investigations* that pain is not a *something*, although, of course, it is not a *nothing* either.[10] To have a pain is not to own anything.

Strawson: [*lighting another cigarette*] I can see that the terminology of ownership is perhaps misleading. But it seems to me that I can abandon this unfortunate turn of phrase, and yet still continue to argue that different people cannot have the same pain – the numerically identical pain – since pains are identity-dependent upon the person who is suffering. We can only refer to a particular pain in so far as we can identify the person who has the pain. We can only identify a particular pain by reference to who has it.

Wittgenstein: Let's pause a moment and see where we've arrived. We know from the Siamese twins that the question of pain-location is really not the crucial issue. We've cast doubt on the idea that having a pain is a relation between a sufferer and a pain. We've repudiated the thought that the subject of pain is the owner of pain. Good. [*He pauses*]

Now, what remains of the claim that my pain must differ from yours – that we cannot have the very same pain? If my pain tallies with yours in phenomenal qualities, such as burning, throbbing, stinging, nagging, if it has the same intensity, and has a corresponding location, what differentiates my pain from yours?

Frege: Wittgenstein, you are just going round in circles. You have come back to the original point I made: You can't have my pain, because mine is mine and yours is yours!

Strawson: Or to put it in the logical mode: pain is identifiability-dependent upon the person who has the pain. That is why different people can of course have an exactly similar pain – that is, a qualitatively identical pain – but not a numerically identical pain.

Wittgenstein: Yes, that is the heart of the matter. What this amounts to is that the property that differentiates my pain from yours is *the property of being mine*. That is, even if there is no difference between my pain and your pain in respect of location, intensity and phenomenal characteristics, still it is a different pain because yours is yours and mine is mine. Someone might even say that by Leibniz's Law that makes it a different pain. But this, in effect, is to treat the subject of pain as *the differentiating property of the pain*. But that is absurd. *I* am not a property of my pain. Nor is *being mine*. A substance is not an identifying property of its properties. Nor is *belonging to a given substance* an identifying property of its properties. Let me explain.

Belonging to me can be said to be a property of my car, precisely because legal ownership is a genuine relation. *Belonging to me* is a relational property of my car, which the car now has, and will lose when I sell it. But, as we have seen, *belonging to me* is not a property of my pain, precisely because *having a pain* is not a relation and a pain is not a relatum.

Look, to say that *being mine* or *being yours* is a relational property of a pain is like saying that an identifying property of the brown colour of your armchair is that it *belongs* to your armchair. No one would want to say that – for then you would have to say that your armchair cannot have the same colour as my armchair. One can't argue that two different objects can't have the same colour because the colour of A belongs to A and the colour of B belongs to B. That is absurd, because it treats the chair – or 'belonging to the chair' – as a differentiating property of its colour.

Searle: Yeah, why not? The colors are different *tokens* of the same color. The token color of this chair is a different token from the color of that chair.

Wittgenstein: No, no. Perhaps we can accept Peirce's distinction between type and token *words* although it is not without its problems. But how could we distinguish type and token colours? If there are three blue patches on the wall and two red patches – there are five different coloured patches, but only two different colours, not five. Types and tokens are just a red herring.

Now, if two objects have a colour of the same hue, intensity and saturation, then they have the same colour. The distinction between being identical and being exactly similar but not the same has no application to colours – or for that matter to weights or lengths.

Strawson: Wittgenstein, I don't quite see. And I can't see that you have shown that I was mistaken to say that pains are identifiability-dependent on the subject of pain. Surely, we cannot identifyingly refer to particular experiences, such as pains, or visual experiences, or emotions, except as the states or experiences of some identified person.

Wittgenstein: Well, of course I can refer to *your* headache and then go on to say that it is getting worse. But do you mean that the phrase 'my pain' or 'your pain' *identifies* a pain?

Strawson: Quite so.

Wittgenstein: But not at all! Does the phrase 'the colour of Smith's chair' identify a colour? When I tell you that I want to buy cushions that are the same colour as Smith's chair, do you know what colour the chair is? The phrase 'The colour of Smith's chair' does not identify a colour at all – and no more does the phrase 'my pain' identify a pain. '*My* pains' – what pains are they? What counts as a criterion of identity here? – Certainly not *being mine*![11] If I tell you I have a pain, you still have no idea *what* pain I have, just as if I tell you that I have a coloured glass goblet, you do not know what colour the goblet is. You don't '*identifyingly* refer to a colour' by the phrase 'its colour'.

Strawson: But surely there is a difference. Material particulars can be identified and referred to without reference to any other kind of entity. That is why I wrote that material objects are basic particulars in our conceptual scheme. By contrast, individual events and processes are dependent particulars – for they can be identifyingly referred to only by reference to the material things – the substances – that are undergoing change. The event or process of *this* chair changing its colour is identified by reference to the chair whose colour fades. It is a different event or process from *that* chair's fading – precisely because they are different chairs, and the processes of fading are identified by reference to the two different chairs. Now isn't it just like that with pains or experiences?

Wittgenstein: Why should it be? A pain is a sensation, not an event. You may have a pain in your tooth, but a pain in your tooth is not an event in your tooth. The *occurrence* of a sharp stab of pain in your tooth might be said to be an event – but *the pain* you have is not an event. If your toothache gets worse, it is not an event that gets worse, and a throbbing toothache is not a throbbing event. After all, the colour of your hair is not an event, even though your hair's changing colour – being dyed for example – is.

Künne: Professor Wittgenstein, I think that you are mistaken. Of course, we say such things as 'Ann and I have the same pain, a throbbing headache in the temples'. But if this were a true identity statement to the effect that Ann's pain is my pain (rather than to the effect that the kind of pain Ann has is the kind of pain I have), then her pain could not have begun before mine, it could not get worse without mine getting worse and an executioner could remove her headache by beheading me.[12] Now, in your *Blue Book* dictation, on pages 54–55, you said that this is no argument. But I don't think that is much of a counter-argument.

Wittgenstein: *Of course it is no argument!* The counter argument is *obvious!* If by 'a true identity statement' you mean an identity statement such as 'This is the same chair as the one you saw in the auction last week – I bought it', then *of course* 'I have the same pain as you' is not what you are calling a true identity statement. For what you mean by that phrase is a statement of what you have been calling *numerical identity.* And to be sure, what I have been urging you all along is that the distinction between being the same and being exactly similar, or, as you and Strawson put it, being numerically and being qualitatively identical cannot intelligibly be applied to pains, or, for that matter, to colours. You are assuming, without any warrant whatsoever, that because 'I have the same pain as you' is not a case of numerical identity, therefore it is a case of qualitative identity. But that is precisely to beg the question.

[*He jumps up excitedly*]

Look, Ann and I have the same colour hair. That does not mean that when Ann dyes her hair red, my hair turns red too – it simply means that we *had* the same colour hair and now we *no longer* have the same colour hair. This armchair [*Wittgenstein thumps on the arm of Künne's armchair*] is the same colour as *that* armchair [*he thumps Strawson's armchair*] – it is not 'numerically' the same colour, nor is it 'qualitatively' the same colour (there is no such distinction here). It is just the same colour – for both chairs are dark brown. Similarly, if Ann and I have the same headache, a dull throbbing headache in the temples, and Ann takes an aspirin which puts an end to her headache, *of course* that does not mean that my headache will also stop. It means that we'll cease to have the same headache.

Künne: Well, I'm not sure … . But let me try another tack. You said that to suggest that pains are individuated by their owners is tantamount to the bizarre claim that the person who suffers is a property of the pain she has. But that can't be right. I do not declare the Earth to be a property of its axis by saying that the axis of the Earth is, just as such, different from that of any other heavenly body. I do not declare Socrates to be a property of his death by saying that the death of Socrates is, just as such, different from the death of any other human being.

Actually, I don't think that you are right to compare pains to chairs at all; you should have compared pains to axes of objects or to deaths of human beings.[13]

Wittgenstein: Come now, Künne. I *contrasted* pains with chairs and compared them with *colours*. But I hope everyone realizes that I am *not* suggesting that the grammar of pain is just like the grammar of colour – it most certainly is not. After all, you can be blue all over [*he smiles*] like the ancient Britons, but you can't be pain all over. But there is an important analogy between having a pain and being of a certain colour – an analogy made visible by the similarity of their grammars with respect to sameness and difference. The fact that this chair was painted green later than that chair does not show that they are different in colour, just as the fact that your headache began later than mine does not show that we do not have the same headache – a mild throbbing pain in the left temples.

Künne: Why shouldn't we consider pains to be in the same boat as grins? Jack's grin is identity-dependent on him. That's why Lewis Carroll's Cheshire cat is funny – for no cat, no smile. So too a person's pain is identity-dependent on the person.

Wittgenstein: Well, Jack's grin lasts as long as Jack is grinning. When he stops grinning he doesn't have a grin on his face any longer. But, of course, Jack's son may have his father's grin. The reason the Cheshire cat amuses us is because there are no grins without grinning faces, no smiles without smiling faces – not because the grinner or smiler supplies us with a criterion of identity for grins and smiles.

Now let me get back to your previous two analogies – between pains and events and between pains and axes of rotation. I have already explained why pains are not comparable to events. I agree that the death of Socrates is a unique event, but Socrates's headache is neither unique nor an event – it is a sensation. I suggested that, in order to shed light on the identity of pains, we should compare the grammar of sensation with that of colour. You suggested that we should compare pains to axes of rotation? Well, is it even true that the axis of rotation of the Earth is, as you put it, *just as such*, different from that of any other heavenly body? What is the axis of the Earth? Do you mean the axial tilt? Well, what is it?

Searle: As far as I remember, it's 23 or 24 degrees.

Wittgenstein: Just so. And suppose that one of the many moons of Jupiter also has an axial tilt of 23 to 24 degrees. Won't it then have the same axis as the Earth? [*Silence*] Or do you mean the *length* of the Earth's axis?

Searle: That's about 7,900 miles.

Wittgenstein: So. There are no other planets in the solar system with *that* axial length – but among the billions of planets scattered throughout the universe, it is more than likely that many will have exactly the same axial length.

Künne, the simple fact is that the phrase 'the axis of rotation of the Earth' does not say *what* axis of rotation the Earth has. No criterion of identity is associated with the phrase 'the axis of the Earth'. But if you tell us that the axial tilt of the Earth is 23 degrees and its axial length is 7,900 miles, then there is no reason why some other heavenly body should not have exactly the same axis of rotation. And to be sure, the axis of the Earth can and does change – and if it does, it will still have *an* axis – indeed *its* axis, but no longer the same axis. And if it had the same axis as the planet Lauflin, then after its tilt diminishes, it will no longer have the same axis as Lauflin.

The phrase 'axis of the Earth' refers to the axis of rotation of the earth, just as the phrase 'A's pain' refers to A's pain – but neither phrase *identifyingly refers*. If you say 'No other heavenly body can have the same axis of rotation as the Earth', we can ask you what the axis of rotation of the Earth *is*, what its tilt and length are – and when you answer, you give us criteria of identity for its axis – and, of course, other heavenly bodies may have the same axis. Similarly, Jack's pain refers to the pain Jack has, but it does not *specify what pain Jack has*. Once you specify the pain as a throbbing headache in the left temple, then we have a criterion of identity, and another person may well have the same pain.

Frege: [*waving his pipe indignantly*] But Wittgenstein, do you really mean to say that you can have my pain? Surely that is nonsense.

Wittgenstein: Well, it depends what you mean. If A has eaten some bad food, he might get a headache. Now B may eat the same food, and half an hour later he may have a headache with exactly the same phenomenal features, and say to A: 'Now I've got your headache'. What he means is that now he has the same headache A had. That is innocuous. It doesn't mean that A's headache has 'migrated' from A's head to B's head. It doesn't imply that A has ceased to have a headache. If someone in such circumstances were to say 'I have your headache', we might well take him to mean that he now has the same headache as you – and that makes perfectly good sense, as long as we don't conceive of *having a pain* as a relational property between a person and a pain. The fact that A's head started to hurt before B's does not mean that A had a different headache than B – only that the event of A's head starting to ache is a different and prior event from B's head staring to ache. So too, your armchair may have become brown before Strawson's did – if the leather on your armchair was dyed before the leather on Strawson's armchair. But that does not show that the two chairs are not exactly the same colour.

Strawson: But Wittgenstein, surely the phrase 'I have your headache' grates – and we are inclined to want to exclude it as senseless. We are all inclined to say, 'You can't have my headache'.

Wittgenstein: Yes. That's all right. You can exclude this misbegotten phrase from currency. But only on two provisos. First, that this does not imply that we can't have *the same* headache. Secondly, that you realize that if *you* can't have my headache, then *I* can't have my headache either.

Frege: *Ach*, Wittgenstein, what on earth do you mean?

Wittgenstein: What I mean is that if the phrase 'I have' does not have the logical multiplicity to accept 'your pain', then it doesn't have the logical multiplicity to accept 'my pain' either. Isn't that obvious?

Frege: No.

Wittgenstein: But it is obvious. Look: what does 'my pain' *mean?* It means 'the pain I have'. But if so, then 'I have my pain' means 'I have the pain I have', which says nothing about my pain (just as 'Either it's raining or it's not raining' says nothing about the weather).

Künne: I'm exhausted.

Strawson: I think we all are. Let's have another glass of this excellent fortified nectar, and go out into the garden and look at the dazzling stars.

Frege: *Ja,* I need a large schnapps after that!

Searle: Wittgenstein, thank you – that was fun. Let's go and watch the night sky.

[*The others all pour themselves fresh drinks and wander out into the garden*]

Wittgenstein [*to himself*] Very interesting. Very interesting. I wonder …
[*He follows them through the French windows with his glass of water*]

Notes

1   Other examples are 'Why can nothing be simultaneously red all over and green all over?', which baffled analytic philosophers in the interwar years, until they gave up; or 'Why can't one imagine what one is currently perceiving?'; or 'How do you know whether you are in pain?'; or 'When one dreams that something is so, does one also believe that it is?'

2   Frege, 'Thoughts' [1918], repr. in *Collected Papers on Mathematics, Logic and Philosophy* (Blackwell, Oxford, 1984), p. 360 [original pagination in German, p. 66].

3   J. S. Mill, *An Examination of Sir William Hamilton's* Philosophy [1865], repr. in Mill's *Collected Works* (University of Toronto Press, Toronto, 1979), vol. ix, p. 192. The emphases are Frege's.

4   J. R. Searle, *The Rediscovery of the Mind* (MIT, Cambridge, Mass., 1994), p. 94.

5   P. F. Strawson, *Individuals* (Methuen, London, 1959), p. 97.

6   See Wittgenstein, *Philosophical Investigations* (Wiley-Blackwell, Oxford, 2009), §254.

7   Frege. *Philosophical and Mathematical Correspondence*, letter to Husserl, 1906 (Blackwell, Oxford, 1980), p. 68.

8   Frege, *Foundations of Arithmetic*, §27: 'Another man's idea is, *ex vi termini*, another idea.'

9   Searle, *Mind: A Brief Introduction* (Oxford University Press, New York, 2004), p. 135.

10   Wittgenstein, *Philosophical Investigations*, §304.

11   Ibid., §253.

12   Wolfgang Künne, 'Wittgenstein and Frege's *Logical Investigations*' in H.-J. Glock and J. Hyman (eds), *Wittgenstein and Analytic Philosophy, Essays for P. M. S. Hacker* (Oxford University Press, Oxford, 2007), p. 39.

13   Ibid.

SUPPLEMENTARY READING

Frege, 'Thoughts' [1918], repr. in *Collected Papers on Mathematics, Logic and Philosophy* (Blackwell, Oxford, 1984).

M. S. Hacker, *Wittgenstein – Meaning and Mind*, vol. 3 of *An Analytic Commentary on Wittgenstein's* Philosophical Investigations. Extensively revised edition (Wiley-Blackwell, Oxford, 2019). Part I, *Essays* – 'Only I Can Gave'; Part II, *Exegesis* §§253–54.

J. S. Mill, *An Examination of Sir William Hamilton's* Philosophy [1865], repr. in Mill's *Collected Works* (University of Toronto Press, Toronto, 1979), vol. ix, p. 192.

P. F. Strawson, *Individuals* (Methuen, London, 1959), p. 97.

L. Wittgenstein, *Philosophical Investigations* (Wiley-Blackwell, Oxford, 2009), §§253–54.

Printed in the USA
CPSIA information can be obtained
at www.ICGtesting.com
JSHW082200140824
68134JS00014B/333